P9-CJN-573

Industrial Gases

Industrial Gases

BY

NORMAN BOOTH
B.Sc., Ph.D., F.R.I.C.

Formerly Chief Executive, Scientific Division,
British Oxygen Co. Ltd.

PERGAMON PRESS
OXFORD · NEW YORK
TORONTO · SYDNEY · BRAUNSCHWEIG

Pergamon Press Ltd., Headington Hill Hall, Oxford
Pergamon Press Inc., Maxwell House, Fairview Park, Elmsford,
New York 10523
Pergamon of Canada Ltd., 207 Queen's Quay West, Toronto 1
Pergamon Press (Aust.) Pty. Ltd., 19a Boundary Street,
Rushcutters Bay, N.S.W. 2011, Australia
Vieweg & Sohn GmbH, Burgplatz 1, Braunschweig

First edition 1973

Library of Congress Cataloging in Publication Data

Booth, Norman.
Industrial gases.

(The Commonwealth and international library. Chemical industry)
Bibliography: p.
1. Gases. I. Title.

TP242.B63 1973   665'.7   72-10293
ISBN 0-08-016860-4

Printed in India by Thomson Press (India) Ltd.

Contents

EDITORS' PREFACE   vii

AUTHOR'S PREFACE   ix

1. Units   1
2. General Information   5
3. Historical Development of Low-temperature Techniques   10
4. Principles of Air Separation   16
5. Distribution of Permanent Gases   26
6. Oxygen   32
7. Nitrogen   48
8. Argon   59
9. Helium   64
10. Other Inert Gases   72
11. Acetylene   75
12. Hydrogen   88
13. Carbon Dioxide   95
14. Miscellaneous Gases   99

GLOSSARY   109

SUPPLIERS OF GASES AND PLANTS   110

BIBLIOGRAPHY   111

INDEX   113

Editors' Preface

WE WERE asked by Sir Robert Robinson, O.M., P.P.R.S., to organize the preparation of a series of monographs as teaching manuals for senior students on the chemical industry, having special reference to the United Kingdom, to be published by Pergamon Press as part of the Commonwealth and International Liberary of Science, Technology, Engineering and Liberal Studies, of which Sir Robert is Chairman of the Honorary Editorial Advisory Board. Apart from the proviso that they were not intended to be reference books or dictionaries, the authors were free to develop their subject in the manner which appeared to them to be most appropriate.

The first problem was to define the chemical industry. Any manufacture in which a chemical change takes place in the material treated might well be classed as "chemical". This definition was obviously too broad as it would include, for example, the production of coal gas and the extraction of metals from their ores; these are not generally regarded as part of the chemical industry. We have used a more restricted but still a very wide definition, following broadly the example set in the special report (now out of print), prepared in 1949 by the Association of British Chemical Manufacturers at the request of the Board of Trade. Within this scope, there will be included monographs on subjects such as coal carbonization products, heavy chemicals, dyestuffs, agricultural chemicals, fine chemicals, medicinal products, explosives, surface active agents, paints and pigments, plastics and man-made fibres.

We wish to acknowledge our indebtedness to Sir Robert Robinson for his wise guidance and to express our sincere appreciation of the encouragement and help which we have received from so many individuals and organizations in the industry, particularly the Association of British Chemical Manufacturers, now merged in the Chemical Industries Association Ltd.

$$\left.\begin{array}{c} \text{J. Davidson Pratt} \\ \text{T. F. West} \end{array}\right\} \textit{Editors}$$

Author's Preface

THERE is no well-established and generally accepted definition of the term Industrial Gases. Broadly this book is concerned with gases which are produced for wide distribution, but excluding the halogens. Some, such as natural gas, synthesis gas, ammonia, town's gas and the products of oil refineries are discussed only so far as is necessary to make the book self-contained.

In dealing with industrial gases special problems arise in connection with measurements, particularly of pressure, temperature, volume and weight. Not only are there several units currently used for each of these parameters but the advent of metrication and the adoption of SI units makes the matter more complicated. An introductory chapter on units has therefore been included and the reader is advised not only to read this first but also to refer to it as often as is required later.

The units adopted throughout the text are those currently in commercial use in the United Kingdom. Factors for conversion to other units are given in Chapter 1.

A short bibliography has been provided to indicate to the reader suitable information sources if he wishes to study some aspects of the subject in greater detail.

CHAPTER 1

Units

THE most important quantities which need to be measured in connection with industrial gases are temperature, time, pressure, volume and weight. In each case there are a number of different ways of expressing the quantity, and practice differs between the United Kingdom, the United States of America and the continent of Europe, although there are some common units. Further United Kingdom practice is gradually changing over a period of years as a result of the decision to make the metric system the primary system for weights and measures from 1975 onwards. The units proposed in the Système International (SI units) are not always the most appropriate in industrial practice and it is uncertain to what extent they will be adopted.

With all these uncertainties it has been thought simplest to use throughout this book the units currently in general use in the United Kingdom. In this chapter alternative units are discussed and some conversion factors given.

TEMPERATURE

The Fahrenheit scale (F), devised by the German scientist Fahrenheit in 1714, takes the freezing point of water to be 32° and the boiling point to be 212°. This has proved a most useful scale in medicine and engineering for the range of temperatures encountered in everyday life. The Centigrade scale (C), suggested by the Swede Celsius in 1742, takes the freezing point of water to be 0° and its boiling point to be 100°.

A more fundamental scale was proposed by the British scientist and engineer Lord Kelvin in 1848. On this scale absolute zero, the lowest possible temperature theoretically achievable, is 0° and the scale units are the same as on the centigrade scale. Since absolute zero on the centigrade scale is $-273 \cdot 15°$ it follows that the freezing point of water on the Kelvin scale (K) is $273 \cdot 15°$.

A further scale used by some engineers, particularly in America, is the Rankine scale on which absolute zero is 0° and the scale units are the same as on the Fahrenheit scale. The freezing point of water on this scale is $491 \cdot 69°$ and its boiling point is $671 \cdot 69°$.

Specialists in cryogenics, the science of low temperatures, tend to use one of the two absolute scales of temperature. In this book, which is intended for the more general reader, the more familiar centigrade scale will be used.

TIME

Rates of gas compression, plant outputs, evaporation losses and so on are frequently quoted somewhat indiscriminately in terms of minutes, hours, days or even months. The SI unit is the second, but this is inconveniently short for industrial purposes. The principal unit used in this book will be the day.

PRESSURE

Most of the pressures referred to in this book are in the range 5 to 200 times atmospheric pressure and a convenient practical unit is the atmosphere, i.e. the pressure equivalent to a column of mercury 760 mm. high at 0°C in a location where the acceleration due to gravity is $980 \cdot 665$ cm/sec². When a gauge reads, say, five atmospheres it means that the pressure is five times atmospheric pressure, so that the total absolute pressure is 6 atm. The pressures quoted in this book are gauge pressures.

In British and American practice a commonly used alternative unit of pressure is pounds per square inch (psi) which is related to the atmosphere by the relationship:

$$1 \text{ atm} = 14 \cdot 695 \text{ lb/in}^2.$$

In continental practice a common unit of pressure is kilogrammes per square centimetre which is related to the atmosphere by the relationship:

$$1 \text{ atm} = 1 \cdot 033 \text{ kg/cm}^2.$$

The SI unit for pressure is Newtons (N) per square metre, the Newton being a fundamental unit of force. This is an inconveniently small unit at pressures above atmospheric and a subsidiary unit is the bar which is 10^5 N/m². It is related to the atmosphere by the relationship

$$1 \text{ atm} = 1 \cdot 0132 \text{ bar.}$$

In some places in this book the pressures quoted are only approximate and in those cases the bar and kg/cm² can be regarded as identical with the atmosphere.

VOLUME

The unit of volume adopted in the United Kingdom and the United States is the cubic foot and this unit has been used throughout this book. On the Continent the unit used is the cubic metre which is also the SI unit. The relationship between these two units is as follows:

$$1 \text{ ft}^3 = 0 \cdot 028317 \text{ m}^3$$
$$1 \text{ m}^3 = 35 \cdot 3148 \text{ ft}^3.$$

It should be noted that in the case of either unit it only has commercial significance if the pressure and temperature at which the volume is measured is standardized. In scientific work it is usually referred to 0°C and 760 mm mercury pressure. In industrial work it is usual to make the reference conditions 60°F (15·5°C) and 30 inches of mercury (762 mm). This latter is the standard adopted in this book.

WEIGHT

Throughout this book the normal United Kingdom standard weights of the pound and the ton containing 2240 pounds have been used. This latter is sometimes referred to as the long ton to distinguish it from the

United States short ton which contains only 2000 pounds. On the Continent the units used are the kilogramme, which is also the SI unit, and the tonne which contains 1000 kilogrammes.

The relationships between these quantities are as follows:

$$
\begin{aligned}
1 \text{ long ton} &= 2240 \text{ lb} = 1016 \cdot 05 \text{ kg} \\
1 \text{ short ton} &= 2000 \text{ lb} = 907 \cdot 18 \text{ kg} \\
1 \text{ tonne} &= 2204 \cdot 6 \text{ lb} = 1000 \text{ kg} \\
1 \text{ kg} &= 2.2046 \text{ lb} \\
1 \text{ lb} &= 0 \cdot 4536 \text{ kg}
\end{aligned}
$$

It will be clear that when approximate figures are quoted, for example in plant outputs, the long ton and the tonne are almost identical.

CONVERSION DATA

In the case of the principal atmospheric gases it is customary to quote moderate quantities in cubic feet and larger quantities in tons. The relationships between these quantities depends on the density of each gas and are as follows:

$$
\begin{aligned}
1 \text{ ton oxygen} &= 26{,}540 \text{ ft}^3 \\
1 \text{ ton nitrogen} &= 30{,}330 \text{ ft}^3 \\
1 \text{ ton argon} &= 21{,}260 \text{ ft}^3
\end{aligned}
$$

the volume in each case being measured at $15 \cdot 5°C$ and 762 mm pressure.

CHAPTER 2

General Information

IN THE gaseous state molecules of matter are sufficiently far apart to be only slightly affected by the attractive forces exerted by other molecules present. Each molecule travels with a constant velocity until it collides either with another molecule or the walls of the containing vessel. Thus, for example, a molecule of nitrogen at room temperature and atmospheric pressure travels on the average at about 1500 feet per second and undergoes about 5000 million collisions per second.

While most substances can be converted to the gaseous state at elevated temperatures, this book is concerned only with those substances whose boiling points are below normal atmospheric temperature and are also articles of commerce. Indeed as mentioned in the Preface even certain materials falling within this definition have been excluded.

It is often necessary to liquefy these gases either in the course of production or for transportation or utilization. In the special case of carbon dioxide it may be transported or used in the solid state. The term "industrial gases" is therefore to be taken to mean these substances whether in the gaseous, liquid or even solid state.

The materials which will be dealt with are shown on the next page in Table 1 together with some of their principal properties.

The first seven materials in Table 1 are ingredients of air in concentrations which are constant near the earth's surface and it is of interest to consider briefly how the composition of the atmosphere was established. It only became realized slowly that different kinds of gas existed, just as more obviously different solid and liquid materials did, and further that different gases had quite different properties. The true nature of the main constituents of air was first established by Lavoisier in 1774 who observed that when he heated tin in air in a closed vessel only part of the

5

air would react. He also made crude quantitative measurements which were gradually improved by other workers. In 1778 Black showed that

TABLE 1. PROPERTIES OF GASES

Gas	Chemical formula	Molecular weight	Boiling point, °C	Melting point, °C	Critical temperature, °C	Critical pressure, atm
Oxygen	O_2	32·0	$-182·97$	$-218·8$	$-118·38$	50·14
Nitrogen	N_2	28·0	$-195·80$	$-210·0$	$-146·95$	33·5
Argon	Ar	39·9	$-185·86$	$-189·4$	$-122·29$	48·34
Neon	Ne	20·2	$-245·9$	$-248·6$	$-228·7$	26·9
Xenon	X	131·3	$-108·12$	$-111·8$	16.6	57·6
Krypton	Kr	83·7	$-153·40$	$-157·2$	$-63·8$	54·2
Helium	He	4·0	$-268·93$	[a]	$-267·97$	2·27
Hydrogen	H_2	2·0	$-252·77$[b]	$-259·2$	$-239·97$[b]	12·98[b]
Ozone	O_3	48·0	$-111·9$	$-192·5$	$-12·1$	54·6
Carbon monoxide	CO	28·0	$-192·0$	$-207·0$	$-139·0$	35
Carbon dioxide	CO_2	44·0	[c]	[c]	31·04	72·85
Ammonia	NH_3	17·0	$-33·35$	$-77·7$	133·0	112·7
Nitrous oxide	N_2O	44·0	$-89·5$	$-102·4$	36·5	71·66
Sulphur dioxide	SO_2	64·1	$-10·0$	$-75·5$	157·12	77·65
Acetylene	C_2H_2	26·0	[d]	[d]	35·5	61·65
Ethylene	C_2H_4	28·0	$-103·71$	$-169·15$	$-9·5$	49·98
Methane	CH_4	16·0	$-161·5$	-184	$-82·1$	45·8
Propane	C_3H_8	44·1	$-44·5$	$-189·9$	95·6	43
n-Butane	C_4H_{10}	58·1	0·6	-135	153	36

[a]Solidifies only at a pressure of at least 25·05 atm at a temperature of $-272°$C.
[b]Figures for 75 per cent ortho. Equilibrium hydrogen, i.e. 99·8 per cent para, has figures of $-252·89°$; $-259·35°$; $-240·26°$; 12·7 atm (see Chapter 12, p. 93).
[c]Solid carbon dioxide sublimes at $-78·5°$C at atmospheric pressure. It melts at $-56·6°$C under a pressure of 5·11 atm.
[d]Solid acetylene sublimes at $-84·0°$C at atmospheric pressure. It melts at $-80·7°$C under a pressure of 1·27 atm.

air contained some "fixed air", now called carbon dioxide, which could be absorbed by lime water. In 1783 Cavendish established, by careful analysis of a number of samples of air, that the atmosphere is substantially constant in composition. His figure of 20·84 per cent by volume for oxygen is remarkably closed to the currently accepted figure. It might be noted that he also showed that a small amount of an inert gas was present in nitrogen obtained from the atmosphere.

During the period 1893–5 Rayleigh observed that the density of nitrogen obtained from the atmosphere by removal of water vapour, carbon dioxide and oxygen was higher than that of nitrogen prepared chemically (14·070 against 14·005 on the scale hydrogen=1). In 1895 Ramsey reacted atmospheric nitrogen with magnesium and found there was a residual completely inert gas which he called argon. Further studies by Ramsey and his colleagues, involving particularly the careful distillation of liquid air, revealed that there were four other equally inert gases present in air, although in considerably lower concentration. The full composition of air is given in Table 2.

TABLE 2.   COMPOSITION OF DRY AIR

Constituent	Meaning	Concentration per cent by volume
Nitrogen	nitre forming	78·08
Oxygen	acid forming	20·95
Argon	no work	0·93
Carbon dioxide		0·03 (variable)
Neon	new	0·00182
Helium	sun	0·00052
Krypton	hidden	0·00011
Xenon	stranger	0·000009

It should be mentioned that the amount of water vapour in the atmosphere is extremely variable. If the temperature is 30°C and the relative humidity 100 per cent, as it may in a warm, wet climate, the water vapour amounts to about 4 per cent by volume. On the other hand, in a cool dry conditions, say at 0°C and a relative humidity of

50 per cent, it is only about 0·3 per cent by volume. This is a point of some importance in the design of air-separation plants.

In addition to the above constituents a wide range of other materials have been reported as present in air in small amounts, depending on local conditions and especially on adjacent industrial plants. These include carbon monoxide, sulphur dioxide, hydrogen sulphide, nitrogen dioxide, methane, acetylene and other hydrocarbons and organic materials. Extremely small amounts of hydrogen and ozone can be detected near ground level; at very great heights the concentrations of these materials are much higher. In the case of ozone this is because the much more intense ultraviolet light from the sun converts oxygen to ozone. In the case of hydrogen it is because very light material tends to accumulate at high altitudes.

The production of any of the constituents of air involves the isolation of the material and not its manufacture in any sense of chemical change. The method of isolation of oxygen or nitrogen might be chemical or physical but in the case of the other constituents, which are completely inert chemically, the method must be a physical one.

While methods involving diffusion, absorption, adsorption or partial crystallization may have some special part to play, the main practical operation is distillation at low temperatures. The reason for the adoption of such an apparently difficult technique is that the difference between the composition of the liquid and vapour phases at any given temperature can be considerable, while other properties may be very similar. A simple example is that if air is completely liquefied the liquid will contain 20·9 per cent of oxygen, but the vapour in equilibrium with this liquid contains only 6·5 per cent.

Where the gases considered in this book are not obtainable from the atmosphere most are deliberately made by chemical processes, i.e. hydrogen, acetylene, carbon monoxide, sulphur dioxide, nitrous oxide, ammonia and ozone. Ethylene, propane and butane are isolated from the products of oil refinery operations while methane is the main constituent of natural gas. Carbon dioxide is either a by-product of fermentation processes or an end product in combustion. Helium is obtained from natural gas.

Although gases can be made with a very high degree of purity, in practice the cost of the later stages of purification tends to be increasingly high. In addition there is usually a small amount of contamination involved in transferring the gas from the production plant to the user, e.g. traces of residual air in cylinders. Fortunately very highly purified gas is not required for most practical purposes. Consequently suppliers of industrial gases operate so as to supply gas only of the purity needed for particular requirements. In discussing each gas in this book reference will be made to the normal purities and to any special purity needed for specific purposes.

CHAPTER 3

Historical Development of Low-temperature Techniques

THE use of low temperatures for storing food dates from Roman times. From then until the Industrial Revolution in Great Britain ice collected during the winter had been used for preserving fish and other perishable foods. Ice rooms were built in a pit in sheltered parts of the garden, the excavated earth being used for thermal insulation.

Deliberate attempts to produce temperatures lower than ambient were not made until the middle of the eighteenth century. When Fahrenheit defined his temperature scale in 1714 he took as the zero point the lowest possible temperature he could obtain by mixing snow and common salt. This is an illustration of a relatively simple but limited method of cooling. When many substances, particularly hydrated salts, are dissolved in water heat is absorbed and the solution becomes cooler than the original materials. The lowest temperature that can be reached for any particular solution is at the eutectic point when the ice and other solid are in equilibrium with the solution. In general the higher the solubility of the substance the lower will be the eutectic point. Table 3 on the next page gives some typical figures.

In 1755 Cullen developed the first ice machine which operated on a different principle. When water is evaporated it needs to be provided with the latent heat of vaporization. If the evaporation is made to take place so rapidly that the residual water cannot absorb heat from its surroundings it becomes cooler and ultimately freezes.

Thereafter many attempts were made to liquefy various gases by a combination of such low temperatures as could be achieved together with high pressures. Particular mention may be made of the work of

10

Faraday who generated his gas in one arm of a sealed tube shaped like an inverted V and immersed the other arm in an appropriate cooling mixture, so that a combination of high pressure and low temperature was achieved in one piece of equipment. Using this technique Faraday was able to liquefy many gases for the first time, including ammonia, chlorine, sulphur dioxide and hydrogen sulphide.

TABLE 3. EUTECTIC MIXTURES

Substance	Per cent by weight in water	Eutectic temperature °C
Sodium chloride	22·4	−21·2
Magnesium chloride	21·6	−33·6
Calcium chloride	29·8	−55
Zinc chloride	51·0	−62

Despite the efforts of other workers who used still higher pressures there remained a number of gases which could not be converted to the liquid form by these techniques and these became known as "permanent" gases. It was only when the concept of critical temperature was developed following Andrew's work on carbon dioxide in 1869 that the liquefaction of the permanent gases became possible. The critical temperature is defined as that temperature above which it is impossible to liquefy a gas however great the applied pressure. The critical temperatures of the principal permanent gases are shown in Table 1 (p. 6). It should be noted that the critical pressure is the pressure required to liquefy the gas at its critical temperature. At lower temperatures a lower pressure suffices.

It will be clear that neither of the methods of cooling substances below the temperature of their surroundings so far mentioned, namely evaporative cooling of water or the use of freezing mixtures, is at all adequate for liquefying the permanent gases. Apart from special methods of academic interest, mainly in attempts to approach as close as possible to absolute zero, there are two practical principles which can be employed. The first of these is to make the gas perform work under such conditions that it cannot absorb much heat from its surroundings. If a gas is compressed it warms up. If compressed gas is allowed to expand into a cylinder it performs work in driving the piston along and cools down. In practice the gas after compression must be brought back to room temperature, before allowing it to drive the piston, otherwise little overall cooling will be obtained. This method of cooling was the subject of patents by Siemens in 1857 and Solvay in 1885 and about the turn of the century was used by Claude in a practical air-separation plant.

The other principle which can be used derives from some observations made by Joule and Thomson in 1852. If a gas under pressure is forced past a constriction such as a felt porous plug, a very small orifice or a partially opened valve, the gas undergoes a temperature change. This effect is referred to as the Joule–Thomson effect and would be zero for a perfect gas. It is observed because any actual gas differs from a perfect gas in two respects. Firstly, there are attractive forces between the gas molecules; when the gas expands work has to be done to overcome these forces and cooling of the gas results. Secondly, in an actual gas the molecules occupy a finite volume which tends to keep them apart; consequently when the gas expands slight warming occurs. The combined result is the Joule–Thomson effect. At temperatures not too much above the condensation temperature the cooling effect predominates. At sufficiently high temperatures the heating effect outweighs the cooling effect. At atmospheric temperature the only gases for which this is observed are helium, neon and hydrogen. The temperature to which a gas must be cooled to obtain a cooling effect on Joule–Thomson expansion is called the inversion temperature.

It will be clear that the Joule–Thomson effect can be used to cool all gases to temperatures below room temperature except helium, hydrogen and neon. In order to make use of the effect in the case of these gases they must be partly cooled by other methods to below their inversion temperatures, i.e. $-3°C$ for neon, $-69°C$ for hydrogen and $-233°C$ for helium. The Joule–Thomson effect was applied in practical air-liquefaction plants about the turn of the century by Hampson in the United Kingdom and von Linde in Germany.

Oxygen was first liquefied in 1877 by two workers quite independently and using quite different methods. Cailletet cooled oxygen to $-29°C$ by a bath of evaporating sulphur dioxide, compressed it to a pressure of

200 atm and released the pressure suddenly. A mist of liquid oxygen was seen. He observed the same phenomenon when using carbon monoxide, acetylene and nitrous oxide but not with hydrogen.

Pictet used a method now known as the "cascade method" which involves proceeding to lower and lower temperatures in a series of stages using different coolants. Thus he liquefied oxygen compressed to 320 atm by cooling it in liquid carbon dioxide at −140°C. The liquid carbon dioxide had in its turn been obtained by cooling in liquid sulphur dioxide.

Liquid hydrogen was first observed by Wroblewski in 1883 when he cooled compressed hydrogen to liquid oxygen temperature and then allowed it to expand rapidly. This is similar to the method used by Cailletet for oxygen, and similarly produced only a mist of the liquefied gas. It was only in 1898 that Dewar collected enough liquid hydrogen for its properties to be measured. Helium, which is even more difficult to liquefy than hydrogen, was first liquefied by Onnes in 1908. His method was to cool compressed helium in liquid hydrogen and then allow it to expand.

It is not possible to give a complete account of all the pioneering work carried out by workers in this field but it should be noted that Wroblewski and Olszewski in the period 1880–90 were the first to obtain relatively large quantities of liquid oxygen, liquid and solid carbon monoxide, liquid and solid nitrogen and liquid ethylene. They employed a cascade method sometimes combined with gas expansion.

Another feature of work with liquefied gases is the difficulty of keeping any product obtained. The influx of heat from the surroundings posed serious problems for the early workers in this field. The major advance was made by Dewar when he perfected a vacuum-jacketed vessel originally designed by d'Arsonval in 1887. Examples of Dewar's containers are available for inspection in the Science Museum in South Kensington, London.

The first commercial plants for oxygen production by the low-temperature distillation method were started up almost simultaneously in England, Germany and France based on technical developments by Hampson, von Linde and Claude respectively. Hampson and von Linde used Joule–Thomson cooling and the countercurrent heat exchangers originally suggest by Siemens in 1857. Claude also used countercurrent heat exchangers but introduced a combined cycle which took advantage of both the high efficiency of the external work principle at higher temperatures and that of the Joule–Thomson principle at lower temperatures. It must be remembered that in addition to their problems in devising the best apparatus for cooling and for heat exchange, they also had to develop equipment for efficient distillation at very low temperatures in order to separate the oxygen from the liquid air.

Since that time there has been steady development in the design of low-temperature separation plants for gases and in methods of storage and transport, particularly in the United Kingdom, the United States, France and Germany. Some of the leading companies are listed on page 110. The principles of air separation will be discussed in more detail in Chapter 4. The whole science of low temperatures (say below −100°C) is referred to as cryogenics from the Greek words *kryos* (icy cold) and *genes* (birth). Indeed the engineering necessary to use low temperatures for practical ends is known as cryogenic engineering. The two together form a considerable field of endeavour, occupying throughout the world the full-time attention of several thousand qualified scientists and their assistants, together with an even greater number of engineers. There are many interesting phenomena which take place at cryogenic temperatures and these will be referred to particularly in Chapters 7 and 9.

The present chapter will be completed with some brief remarks about methods of measuring low temperatures in practice.†

Since mercury freezes at −38·9°C the ordinary mercury in glass thermometer cannot be used in cryogenic work. A general instrument suitable for use from room temperature down to −270°C is the constant-volume gas thermometer. It consists of a pressure-measuring device such as a Bourdon tube connected to a thermometer bulb by a capillary tube, the whole being filled with a low boiling gas, e.g. helium. The gas is kept at constant volume and its pressure, which is proportional to the absolute temperature, is measured. At temperatures below the critical point of the gas the device can act as a vapour pressure thermometer, i.e. one that measures the vapour pressure of a liquid whose vapour

† For a discussion of absolute measurements and thermodynamic ideas of temperature scales at low temperatures the reader is referred to chapter 3 of reference 1 in the Bibliography.

pressure–temperature relationship is known. It will be evident that temperature changes along the capillary tube, which is not immersed in the cryogenic fluid, can introduce errors into the measurement unless a compensating device is incorporated. Further, it is not strictly true in the case of any actual gas to regard its pressure–temperature relationship as that of an "ideal" gas. Consequently there are a number of possible sources of error and a number of corrections to be made in using a gas thermometer. However, with care in this direction, it can give very accurate temperature readings in skilled hands.

In industrial plants, where robustness and reproducibility are more important than absolute accuracy, resistance thermometers are commonly used. Such an instrument depends on the variation in the electrical resistance of a metal with temperature and can be used over a wide temperature range. Thus a platinum resistance thermometer can be used down to $-259°C$, while lead can be used down to $-264°C$ and indium to $-269.8°C$. This type of instrument is relatively cheap and simple to use while at the same time being sensitive, reproducible and robust.

Thermocouples are devices in which pairs of dissimilar metals generate a voltage when one junction is kept at a constant temperature and the other is at the temperature to be measured. They are less accurate than resistance thermometers and tend to become less sensitive at lower temperatures. They are, however, sometimes used in low-temperature work; it is usual to maintain the constant-temperature junction at liquid nitrogen temperature.

CHAPTER 4

Principles of Air Separation

BOYLE'S Law states that the volume of a gas is inversely proportional to its pressure at constant temperature. Charles' Law states that the volume of a gas is proportional to its absolute temperature at constant pressure. The two laws together lead to the relationship $PV = RT$, where P is the pressure, V the volume, T the absolute temperature and R a constant. A perfect gas is defined as one which strictly obeys this relationship.

It is quite a simple matter to calculate the theoretical minimum amount of energy required to separate a mixture of two perfect gases. Since mixing of the two gases results in an increase in entropy, then in separating them work must be done to decrease the entropy by the same amount. The calculation can be illustrated in the following way.

Let us assume that air consists only of 21 per cent of oxygen and 79 per cent of nitrogen and that each behaves as a perfect gas. As we have seen the former assumption is not quite true. The latter is, however, very nearly true since the heat of mixing of oxygen and nitrogen is negligibly small. Imagine that the mixture at atmospheric pressure is contained in a cylinder which has two semipermeable membranes, one permeable to oxygen and the other to nitrogen. Diffusion will occur until all the nitrogen has passed through one membrane and is contained at its partial pressure of 0·79 atm; and all the oxygen is contained at the other end at its partial pressure of 0·21 atm. So far no work has been done on the system, but in order to complete the separation each gas must be compressed isothermally and reversibly to atmospheric pressure. For the nitrogen the work which has to be done is $0·79\,RT \log_e 1·0/0·79$. For the oxygen it is $0·21\,RT \log_e 1·0/0·21$. The total work of separation is the sum of these quantities.†

† For a more rigid account see chapter 6 of reference 3 in the Bibliography.

16

It should be noted that this theoretical figure is quite independent of the practical method of separation adopted. In practice no process can be truly reversible and a major problem in all gas separation processes is to devise a method which will be as nearly reversible as possible, i.e. which has the highest possible thermal efficiency. Far and away the two largest items in the cost of air separation are power and capital. Attention to efficiency is directed towards making the power consumption as low as possible. However, at some stage it is found that the power consumption can only be further reduced at the expense of increased capital expenditure and the further one goes the greater the increase in capital expenditure per unit of power saved. Consequently industrial plants are operated neither at the highest achievable efficiency nor at the lowest possible capital cost, but at a combined economic optimum which lies between these two extremes.

As mentioned in Chapter 2, although methods other than liquefaction and distillation have been considered and carefully studied, none compares with this method in economic terms. The thermodynamic efficiency,† i.e. the ratio of the minimum theoretical amount of energy required to that actually used, does not exceed about 15 per cent for a modern medium-sized plant producing reasonably pure gaseous products. The reasons for this apparently low efficiency will be discussed towards the end of this chapter.

Air-separation plants are built in a wide variety of outputs; to produce one or more components; to separate one or more of the products in liquid form; and to make products of various purities. The precise practical method employed depends to some extent on the definition of these requirements. An air-separation plant, however, always has three basic stages, namely purification, refrigeration and distillation. The other technique of prime importance is that of heat transfer, so as to keep losses of cold to a minimum. Each of these four will now be considered in detail.

---

† The reader should note that the thermodynamic efficiency takes no account of process or cycle. There are also in use two other measures of efficiency. The first is the "process efficiency" which relates the minimum theoretical energy to that theoretically achievable for a particular cycle. The second is the "practical efficiency" which relates the minimum work theoretically possible for a particular cycle to that actually achieved. We shall use only the thermodynamic efficiency in this book.

PURIFICATION

As was seen in Chapter 2 air is a complex mixture of elements together with carbon dioxide and water vapour. Either of these latter materials would solidify in a low-temperature plant and must be removed before the air is further treated. There are a number of methods of carrying out the purification in practice. In small gas plants and in liquid plants using high pressures the air is purified before it enters the separation unit. The carbon dioxide is usually scrubbed out with caustic soda solution at about 10 atm pressure, i.e. after the second stage of the air compressor. Moisture is commonly adsorbed on silica gel or on activated alumina. Units for doing this are provided in duplicate, so that while one is on stream the other is being reactivated, e.g. at a temperature of 300°C with dry nitrogen from the plant.

On really large gas plants operating at relatively low air pressures, say 6 atm, purification can be effected by means of regenerators (Fig. 4.1).

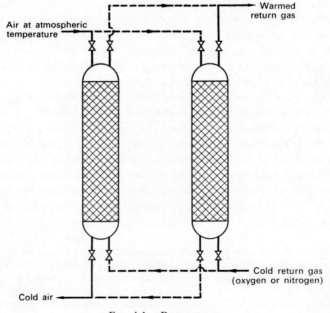

Fig. 4.1.   Regenerators.

The principle involved is to pass the incoming air through a packed vessel already cooled down by cold product leaving the plant. The water vapour and carbon dioxide are deposited on the packing material in the solid state. On reversing the cycle so that the cold plant products pass through the packing to cool it down again, these solid impurities are evaporated from the packing and pass out of the regenerators and the plant. This method naturally results in the plant products being contaminated with the impurities but for some large-scale uses this does not matter.

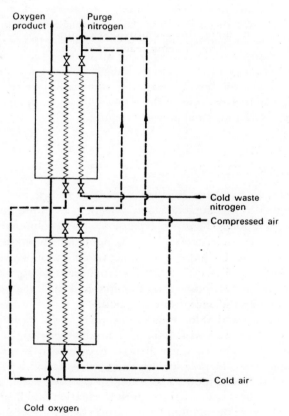

Fig. 4.2. Reversing exchangers.

In the operation of regenerators it is an advantage that the volume of the returning gas is greater than that of the incoming gas because it is at a lower pressure. On the other hand, the temperature of the return gas at any given point in the regenerator is lower than that of the incoming air, which means that its saturation capacity for water vapour and carbon dioxide is lower. However, provided a certain maximum temperature difference is not exceeded complete re-evaporation can be achieved.

An alternative method which avoids the contamination of the products, or at any rate of one product, is to use reversing exchangers. Fig. 4.2 illustrates the use of reversing exchangers when pure oxygen is required. The oxygen leaving the plant passes continuously through the heat exchanger while the incoming air and outgoing nitrogen are switched alternately through the exchanger in order to condense and resublime the carbon dioxide and water vapour. Reversing exchangers have a slightly higher pressure drop through the equipment with the result that a little additional power is required.

REFRIGERATION

Once the air has been purified it can be cooled and ultimately liquefied either by making it do external work in an expansion machine, or by employing Joule–Thomson expansion, or by a combination of both methods. The cascade process (p. 13), although thermodynamically attractive, is not used in air separation because of the difficulty of providing abundant supplies of intermediate materials of suitable boiling points, the number of compressors required and the danger of product contamination by leakage within the plant.

It will be realized that for a perfect gas mixture entering the plant at room temperature and with separated products leaving at the same temperature the only work which need be provided is the reversible work of separation. In practice, however, refrigeration is required not only to cool the plant down and establish a steady state, but also to compensate for heat leakage through the insulation and heat losses due to products leaving the plant at temperatures below atmospheric. If one of the products is withdrawn as liquid then still more refrigeration is required as the product concerned has both sensible and latent cold, i.e. not only the cold due to its low temperature but also the cold due to the latent heat of condensation.

The essential stages involved use both an expansion machine (isentropic expansion) and the Joule–Thomson effect (isenthalpic expansion) as shown in Fig. 4.3. The purified air is compressed to an appropriate

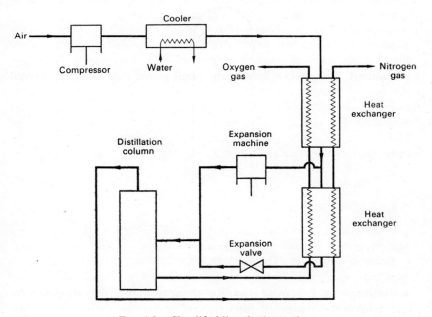

FIG. 4.3. Simplified liquefaction cycle.

pressure and is brought back to room temperature by cooling with water. It then exchanges its heat with outgoing cold products and a portion of it passes to an expansion engine. The work it is compelled to do in the engine cools it further and the partially or completely liquefied air passes to the fractionating column. The remainder of the air passes through a second exchanger and an expansion valve and then goes to the column.

RECTIFICATION

For simplicity let us imagine air to be a simple two-component system of oxygen which boils at about −183°C at atmospheric pressure and nitrogen which boils at −196°C at this pressure. If liquid air is passed to a distillation column (Fig. 4.4) it can be separated into

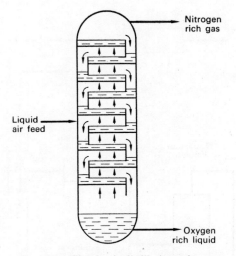

FIG. 4.4.   Simple air-distillation column.

nitrogen-rich gas leaving the top of the column and oxygen-rich liquid leaving the base. However, in order to obtain reasonably pure products, it is essential to maintain a temperature difference between top and bottom of the column and also to provide a reflux at the top of reasonably pure nitrogen liquid. Since the nitrogen gas leaving the column is the coldest available material it will not be possible to provide the reflux in a simple column.

The method normally adopted in practice to deal with this problem is to employ a device developed by von Linde in 1910 and known as a "double column". It is illustrated in Fig. 4.5 and, as will be seen, consists of two interlocking columns, one above the other. The lower column operates at a pressure of about 5 or 6 atm and the upper one at just above atmospheric pressure. In the lower column the feed is separated into pure nitrogen, which at the pressure of 5–6 atm liquefies at −178°C at the top of the column, and an "enriched air" containing 30–40 per cent of oxygen at the base. Each stream after leaving the lower column passes through an expansion valve to a pressure just above atmospheric

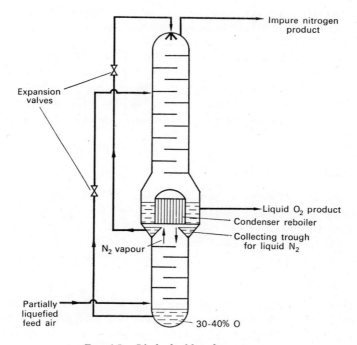

Fig. 4.5. Linde double column.

and in doing so is further cooled. The cold nitrogen liquid enters the top of the upper column and provides the reflux. The cold enriched air enters the upper column at a point part way up and is separated into nitrogen gas and liquid oxygen. Since the latter accumulates at the base of the column at about −183°C it is able to act as a condenser for the nitrogen at the top of the lower column.

Although the design of this double column may appear to be rather complex there are some factors that enable the column to be relatively simple. Because the liquid being handled is extremely clean the trays can be made of perforated plates instead of the bubble-cap trays normally used in other industries. In addition liquid air has a low viscosity and does not foam badly. Consequently tray spacing can be relatively close so enabling the column to be kept as short as possible. In view of the problems of insulating columns at these low temperatures this has a distinct advantage in reducing heat influx.

It will be realized that air is a more complex mixture than has been assumed in this discussion. In particular it contains nearly 1 per cent of argon which boils at a temperature intermediate between oxygen and nitrogen. There are also minor problems arising from the presence of the other inert gases. These points will be discussed in later chapters.

HEAT TRANSFER

Heat-transfer problems arise in two ways on low-temperature plants. Firstly, in order to recover the maximum amount of cold from the outgoing products, transfer of cold to incoming warmer materials must be as complete as possible. Secondly, in order to keep cold losses to the surroundings as low as possible, insulation of the plant must be as good as possible.

Reference was made earlier in this chapter to the use of regenerators and reversing exchangers on large gas plants for the dual purpose of removing impurities and providing good heat exchange. On small gas plants and liquid plants it has been common practice to use a type of exchanger which operates on the crossflow principle. The incoming high-pressure air passes through a series of tubes wound spirally round a central mandril. The low-pressure product leaving the plant passes across the outside of these tubes approximately at right angles.

As far as insulation is concerned it is common practice to enclose all the cold components of the plant in a "cold box" filled with slag wool or an expanded mineral type of insulant. For safety reasons insulants should be non-flammable. Organic material accidently soaked in liquid oxygen can be highly explosive (cf. p. 46).

MATERIALS OF CONSTRUCTION

Mild steel tends to become brittle below about −40°C and is then liable to fracture if subjected to mechanical shock. For parts of a plant which are required to operate at low temperatures it is therefore essential to employ materials such as copper, aluminium or austenitic steels which do not suffer from this defect. The precise choice of material will depend on fabrication requirements and price.

PLANT EFFICIENCY

Reference was made on p. 17 to the low efficiency of air-separation plants compared with the thermodynamic minimum energy required. This inefficiency arises from a number of causes of which the chief are:

1. The air compressor itself departs considerably from thermodynamic reversibility. Efficiencies vary, depending on the design of compressor, its size and the duty it is required to perform. Under the best conditions it rarely exceeds 75 per cent.
2. The distillation stage is accompanied by an increase in entropy.
3. Because the gas has to be passed through heat exchangers at high speed and with appreciable pressure drop there is a rise in entropy.
4. Like a compressor an expansion machine departs from reversibility and there is an increase in entropy.
5. Cold is lost with the products.
6. There is heat influx into the plant from the surroundings.

CHAPTER 5

Distribution of Permanent Gases

THE design of an air-separation plant depends on the products it is required to make and on the method by which they will be conveyed from the plant to the point of use. This in turn depends on the use to which the product will be put, the quantity required, whether the demand is intermittent or continuous and the distance between the production site and the point of use. The distribution of other permanent gases such as hydrogen and helium follows a similar pattern. Methods of distributing other gases dealt with in this book will be described in each relevant chapter. There are three methods commonly used for distributing the permanent gases. These are as gas in cylinders under pressure, as gas through pipelines and as liquid.

CYLINDER DISTRIBUTION

Supplies of small quantities of gas have traditionally been made in mild steel cylinders which could be left at customer's sites and collected when empty. In the United Kingdom the pressure in full cylinders has normally been 132 atm but recently cylinders have been introduced which can be charged to 160 atm. A variety of cylinder sizes has been employed, the capacity being defined as the volume of free gas measured at 15·5°C and 762 mm pressure. The commonest size has a capacity of 240 ft³ of free gas and weighs about 140 lb when empty and 160 lb when full of oxygen at 132 atm. It will be seen that the ratio of total weight to weight of oxygen carried is 8:1. In the case of hydrogen, owing to its lower density the ratio is about 110:1.

Despite the fact that the ratio of weight of metal to weight of gas is so high it is not economic to transport hydrogen in liquid form, partly because of the high cost of making the liquid and partly because of the high evaporation losses incurred in handling such a low boiling liquid. Frequently hydrogen batteries of up to $6 \times 11,500$ ft^3 or $14 \times 3,700$ ft^3 cylinders are transported on a trailer and left at the customer's works, when large amounts of hydrogen gas are required. The modern tendency is to use larger batteries of smaller cylinders, e.g. 96–176 cylinders each holding about 460 ft.

Although the ratio of weight of metal to weight of gas transported is high with all gases, cylinders have a number of operational advantages. The gas can be stored indefinitely without loss and the cylinders can be moved from point to point on a site depending on where the gas is required. However, when demand is high and steady it is often more economic to transport the material in the liquid form.

It should be added that the inert gases neon, krypton and xenon are sometimes transported in low-pressure cylinders or even glass ampoules if only small amounts are required.

In order to avoid confusion cylinders are painted a standard colour or combination of colours for each gas.† Some of the commoner gas cylinders are:

oxygen	black	nitrous oxide	French blue
nitrogen	grey/black	neon	brown/black
argon	blue	methane	red
helium	brown	ethylene	purple/red
hydrogen	red	carbon monoxide	white/red
acetylene	maroon		

It will be realized that in the case of gas mixtures the colouring tends to get more complex.

LIQUID DISTRIBUTION

Distribution of oxygen in liquid form has been an industrial operation in the United Kingdom since 1934. Today there is also considerable movement of nitrogen and argon in this way, and on a smaller scale of

† See reference 5 in the Bibliography for full details.

helium. Distribution in liquid form is carried out either because the product will be used as liquid, or because it is more economic even though it will be converted to gas before use. The relative volumes of a ton of gas and of liquid are given in Table 4.

TABLE 4.   VOLUME OF GAS PER TON

Substance	Volume of gas at 15·5°C and 762 mm pressure, ft^3	Volume of liquid at the boiling point, ft^3
Oxygen	26,540	31·1
Nitrogen	30,330	44·4
Argon	21,260	25·8

It will be seen that there is a considerable saving in volume, even if the gas is compressed to 160 atm. In addition there is a large saving in weight since a liquid transport vessel weighs about half the weight of liquid carried against eight times in the case of gas. Consequently there is about five times as much oxygen carried in liquid form compared with gas for a given total weight or "payload".

The technique employed is to transfer the liquid under gravity or by pump from a storage tank at the production centre into a tank on a lorry, or even on a train. On arrival at the customer's works the liquid is pumped into a storage tank, usually one in which it will later be evaporated to gas. All these vessels must be efficiently insulated in order to reduce heat influx. The two main techniques employed are to insulate with slag wool or similar material or to fill the insulating space with a powder and apply a vacuum. Those parts of the vessel in contact with cold liquid cannot be made of mild steel, but outer vessels which are approximately at room temperature can be.

Large storage tanks at production sites may hold as much as 1,500 tons of liquid oxygen, equivalent to about 40 million ft^3 of gas. Losses in such a vessel may be as low as 0·1 to 0·2 per cent per day. Any gas which evaporates can be collected for compression into cylinders or return to the plant for reliquefaction. A common size of transport vessel holds 11 tons of liquid, although some now hold as much as 30 tons. Evaporation losses are naturally rather greater in transport vessels than in large storage tanks but more serious losses occur during transfer of the liquid from tank to tank. If a transport vessel has been allowed to warm up at any time the initial cooling down will involve a considerable loss of the liquid. There are also losses involved in cooling down transfer lines, hoses and pumps. Distribution in bulk as liquid involves sophisticated techniques in the construction of vessels and equipment, special attention being paid to keeping losses as low as possible. These losses, however, can still amount to about 15 per cent in a well-designed distribution system and are often higher in practice.

At the customer's works the gas is evaporated, the necessary heat being supplied by the surrounding air, or more commonly by hot water or steam. If the gas is required under pressure it may be compressed after evaporation or the liquid may be compressed before evaporation. Alternatively it can be evaporated in a closed vessel, so generating its own pressure. Despite the higher cost of liquid oxygen than gas, the distribution losses and the cost of the special equipment, these are more than counterbalanced by the reduction in transport costs.

As will be seen later nitrogen and helium are sometimes required for use in liquid form. The quantities are frequently small compared with those discussed above and frequently vessels holding 10 to 200 litres are employed. They are delivered full to the customer's works and collected again when empty. The vessels used for nitrogen are generally vacuum-jacketed and constructed in copper, aluminium or alloy steel. Evaporation losses are of the order of 3 per cent per day for a 25-litre vessel and 1·75 per cent for a 75-litre one.

In the case of liquid helium storage or transport not only is a vacuum jacket employed but also a bath of liquid nitrogen which is also vacuum-jacketed (Fig. 5.1). A more recent technique, which can obviate the nitrogen jacket, is to use multilayer reflectors in a vacuum for insulation. An example is closely packed layers of aluminium separated by spaces of fabric. The former reduces heat transfer by radiation and the latter heat transfer by conduction.

Typical evaporation losses in a 25-litre nitrogen jacketed vessel are

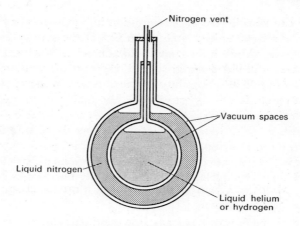

FIG. 5.1.   Small liquid-helium storage vessel.

1·5 per cent per day against 5·0 per cent using the reflecting layer technique. On a scale of 100 litres the losses in the latter case fall to below 1·5 per cent per day.

PIPELINE DISTRIBUTION

A typical lorry with a capacity of 11 tons of liquid oxygen carries the equivalent of about 300,000 ft³ of gas. There are in the United Kingdom a number of steelworks which require quantities of oxygen in excess of 5,000,000 ft³ per day, the biggest demand being about 36 million. It would obviously be a very considerable operation to deliver all this oxygen in liquid form. Moreover, the cost of the product would be too high to justify its use in steel-making processes. It has become essential, therefore, to build large gaseous oxygen plants adjacent to steel-making plants and pipe the oxygen to the points of use. In the United Kingdom the pattern of development has been for one of the companies specializing in oxygen supply to build and operate the plant on behalf of one or more steelworks. For example, at Carfin near Motherwell 1000 tons per day of oxygen can be piped down 15 miles of pipeline to six works. The largest supply on one site is 1000 tons per day at the Steel Company of Wales.

Even with these schemes the oxygen has usually to be piped for several miles to get at all the usage points. Mild steel is used as the constructional material but it must be remembered that this metal will begin to burn in oxygen quite vigorously if the temperature is raised locally to 1250°C, and will continue burning. It is essential, therefore, to ensure that such a temperature cannot be reached locally either by friction or other causes. All pipelines must be free from adventitious materials, scale and oil or grease, and also free from internal projections at the joints.

Although the main gas supplied by pipeline is oxygen there are examples of the distribution of nitrogen by this method, e.g. under the Tees from Middlesborough to Billingham at a rate of about 700 tons per day.

CHAPTER 6

Oxygen

HISTORICAL

Before the development of the low-temperature distillation route oxygen was made by chemical methods. In the United Kingdom Brin's process was operated commercially from 1886 to 1906. It depended on laboratory observations that barium oxide would react with atmospheric oxygen at about 500°C, and that the peroxide so formed would decompose if the temperature was raised to about 800°C.

$$2BaO + O_2 \xrightarrow{\;500°C\;} 2BaO_2 \xrightarrow{\;800°C\;} 2BaO + O_2$$

In commercial practice it was found that the barium oxide lost its power to react after a few cycles unless the incoming air was first freed from carbon dioxide, organic matter and dust. It was also found that, instead of the expensive and time-consuming practice of raising and lowering the temperature by 300°C, the process could be operated at about 600°C using two different pressures:

$$2BaO + O_2 \xrightarrow{\;2\text{ atm}\;} 2BaO_2 \xrightarrow{\;0.05\text{ atm}\;} 2BaO + O_2$$

The gas produced by this process was rather impure by modern standards, usually containing 90–96 per cent oxygen together with nitrogen and smaller amounts of argon and other gases.

From 1906 until 1934, when liquid transport was introduced in the United Kingdom, low-temperature oxygen plants were designed to make gaseous oxygen. By modern standards the scale of operation was small, a typical size being 1000 ft³ per hour, equivalent to slightly below 1 ton per day.

From 1934 plants making liquid have been employed, a typical scale of operation in recent years being 50 tons per day, although plants are in operation which produce as much as 100 tons per day and still larger plants are under construction. Some oxygen can be withdrawn as gas from such plants. The purity of normal commercial oxygen is at least 99·5 per cent.†

After World War II plants were introduced for making gas on a much larger scale, but often of slightly lower purity, for distribution to the point of usage by pipeline. Plants making 200–1000 tons per day are common today and one has been built in Belgium to produce 1500 tons per day. Because the main uses of "tonnage" oxygen have been in the manufacture of steel, the gasification of coal and the oxidation of natural gas, it has been possible to relax the high purity achieved when liquid oxygen is made and the term "medium purity" oxygen has been introduced for oxygen of 95–98 per cent purity. There has, however, been an increasing tendency in recent years to require higher purities for most applications.

The rapid growth of oxygen demand over the last thirty years is illustrated by the following United Kingdom consumption figures:

1938	36,800 tons
1948	126,000 tons
1954	210,000 tons
1959	552,000 tons
1963	1,370,000 tons
1972	1,840,000 tons

The 1972 output was made on about seventy plants and about 88 per cent of it was tonnage oxygen.

There are a wide number of designs of oxygen plants and for purposes of illustration one liquid plant and one large gas plant typical of those used in the United Kingdom will be described.

LIQUID-OXYGEN PLANT

A typical liquid-oxygen plant is shown diagrammatically in Fig. 6.1. This plant operates on the Heylandt cycle, named after the engineer

† See reference 6 in the Bibliography.

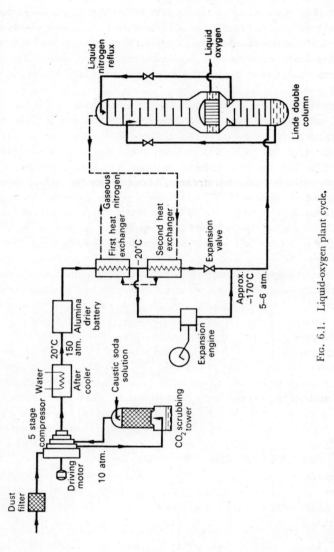

FIG. 6.1. Liquid-oxygen plant cycle.

who first developed it. Air, freed from dust by filtration, is compressed to 150–200 atm in a reciprocating machine having five or six stages. Between each stage and after the last stage the air is cooled by water to remove the heat generated during compression. In between the second and third compression stages the air, which is then at a pressure of about 10 atm, is passed through a caustic soda scrubber to remove carbon dioxide. Much of the water vapour in the original air is condensed out during compression and the remainder is removed by passing the compressed air through beds of silica gel or alumina pellets.

The dry CO_2-free air passes through the first heat exchanger where it is cooled to about $-20°C$ by the returning gaseous nitrogen stream, and is then divided into two streams. About 40 per cent is further cooled by nitrogen to about $-170°C$ and expanded through a valve to reduce the pressure to 5–6 atmospheres. As a result of this operation much of the air is liquefied. The other 60 per cent of the air is expanded to the same pressure through a reciprocating expansion engine. The two streams are combined and fed to the lower part of the double column described in Chapter 4. The reason for dividing the stream is to recover the maximum amount of refrigeration from the nitrogen gas. Although cooling by means of expansion machines is more efficient than through valves the best overall result is the combination described.

The liquid oxygen produced on this type of plant is usually at least 99·7 per cent pure, the balance being argon. The waste nitrogen from the plant contains 2–3 per cent oxygen. The thermodynamic efficiency is about 24 per cent, compared with 15 per cent for medium-sized gas plants and 18 per cent for large gas plants.

LARGE GASEOUS OXYGEN PLANT

When large quantities of gaseous oxygen are required on one site by the steel or chemical industries, opportunities for cost reduction occur which do not occur when higher-purity liquid oxygen is being made. Since the oxygen will leave the plant at approximately atmospheric pressure, the only refrigeration required once the plant is in operation arises from inefficiencies in the heat-transfer system. If perfect heat transfer could be achieved the outgoing cold gas would cool the incoming air down to a suitable temperature for liquefaction prior to entering the distillation column.

The use of regenerators for purification on large gas plants was described on page 18. A regenerator is a vessel packed with a suitable material, such as graded stones, and in essence acts as a cold storer. The regenerators are used in pairs, the cold outgoing gas passing through one for 3 to 10 minutes to cool it, while incoming air is passing through the other one and warming it. At the end of this time the passage of gas through the pair is automatically reversed. Owing to the difficulty of sweeping out air from the regenerator on reversing, it will be seen that the oxygen product will be slightly less pure than when it left the distillation column. In addition as the incoming air is compressed while the outgoing air is only slightly above atmospheric pressure, it is necessary to blow down from higher to lower pressure on reversing the regenerators and this entails slight air loss. Further, there has to be one pair of regenerators for the oxygen product and another for the nitrogen, and in each case the product is contaminated by the evaporated water vapour and carbon dioxide. Nevertheless, the regenerator system provides very efficient heat transfer and has enabled cycles to be developed operating at relatively low pressures and therefore a low power consumption. A large number of plant designs exist because the plant is frequently tailored to precise customer requirements. Figure 6.2 shows a typical design for a plant producing 90–99·5 per cent oxygen containing moisture and carbon dioxide.

The incoming air is passed through a filter and compressed by a turbo-compressor to 5·2 atm gauge. It is cooled and washed by water and then passes through the regenerators. The bulk of the cold air enters the lower fractionating column where it is separated into pure liquid nitrogen at the top and an oxygen-rich liquid at the base. The former passes through a sub-cooler and is expanded through a valve to provide the reflux for the upper column. The oxygen-rich fraction passes through silica gel to remove any traces of carbon dioxide or hydrocarbons and also a sub-cooler before being expanded through a valve into the upper column. The sub-coolers are units for exchange of cold with the nitrogen product leaving the plant.

The remainder of the air is removed from a point part way up the regenerators and is passed through a purifier, expanded in a turbine and

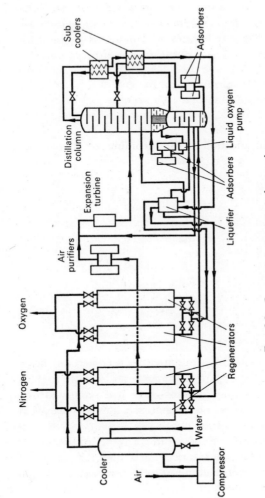

Fig. 6.2. Large gaseous oxygen plant cycle.

passes directly to the upper column. This procedure provides enough refrigeration to compensate for heat leakage and also heat losses in the regenerators.

If the oxygen is required free from carbon dioxide and water vapour then the outgoing oxygen passes through a reversing exchanger (Chapter 4), while the nitrogen continues to pass through a regenerator.

PROPERTIES OF OXYGEN

Liquid oxygen is a pale-blue transparent and very mobile liquid. One volume at its boiling point will evaporate to give 843 volumes of gas at atmospheric pressure and temperature. It is a non-conductor of electricity but it has strong paramagnetic properties; a suspended test tube containing liquid oxygen will move readily towards a magnet.

Combustion proceeds much more actively in oxygen than in air, and combustion processes which are normally under control are liable to become uncontrollable if the oxygen content of the air is raised to as little as 30 per cent. It is therefore possible for a small leakage of oxygen into a room to increase the oxygen level to a dangerous figure. For this reason both liquid oxygen and oxygen cylinders should be stored well away from flammable material and from sources of ignition such as heat, fires and sparks. Oil must not be used to lubricate the valves of oxygen cylinders and all pipelines and connections should be freed from oil and grease before use.

USES OF OXYGEN

For many oxidation processes air, which is really a form of dilute oxygen, is quite satisfactory. However, a great deal of the heat produced during combustion is dissipated in warming up the inert nitrogen, with the result that a lower temperature is achieved than if pure oxygen were used. In some cases this temperature difference is of fundamental importance. Oxygen is used in some instances in place of air because the nitrogen would undergo undesirable chemical reactions. It is also used in some chemical processes because by eliminating the diluent nitrogen the reaction space required can be greatly reduced, and also because the product can be recovered more efficiently, as for example in the oxidation of ammonia to nitric oxide. The most important and wide-spread uses of oxygen are, however, in the metallurgical and engineering industries.

One of the earliest commercial uses of oxygen was to burn acetylene to produce a very high-temperature flame which could be used to weld two metal surfaces together. A fuller account of this process is given in Chapter 11. While oxy-acetylene welding is widely employed, especially for welding steel, it should be noted that the oxy-hydrogen flame is sometimes used for such relatively low-temperature operations as welding aluminium and other low melting-point metals, and also for brazing. Indeed other fuel gases, such as town's gas, natural gas, propane or butane, can be used for certain processes such as the welding of lead.

Some of the main uses of oxygen are described in detail in the following sections. It will be seen that, while most of them require the oxygen in the gaseous form, two—explosives and rocketry—require it in the liquid form.

CUTTING OF METALS

Most metals when heated in oxygen to a sufficient temperature react with it. Usually the oxide has a higher melting point than the metal and the solid oxide tends to hinder further reaction. In the case of iron the oxide formed has a lower melting point than the metal and is easily

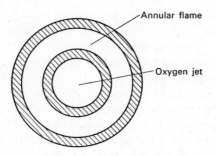

Fig. 6.3. Section of cutting torch.

removed from the surface in liquid form either under gravity or the force provided by a stream of gas. This enables a practical steel-cutting process of reasonable precision to be performed with the right equipment.

A torch is used which consists of a central oxygen stream surrounded by an annular flame as shown in Fig. 6.3. This provides a stream of hot oxygen gas which can be applied to one edge of a steel plate as shown in Fig. 6.4. The metal oxidizes and partly falls under gravity

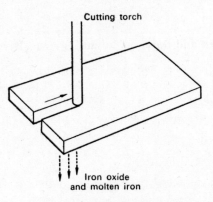

Fig. 6.4.   Cutting of steel.

and is partly blown away by the stream of gas. The torch is moved relative to the metal in the direction in which it is required to effect the cut. The profile of the cut is relatively smooth and clean and can be used without further treatment in other fabrication processes.

It should be noted that the steel needs to be heated only to about 1250°C for reaction with oxygen to commence. Once the exothermic cutting operation has begun the only heat which need be supplied is to compensate for heat lost, mainly by conduction through the metal. The oxy-acetylene flame has a temperature of about 3100°C, which is higher than obtained with any other fuel. The only advantage this gives is the more rapid rate of initial heating so enabling the cutting operation to be started more quickly. In practice it is often more economic to use a cheaper alternative fuel such as propane whose flame temperature in oxygen is about 2800°C. It is of interest to note that the inventor of the cutting process, Fletcher in 1889, used coal gas as his fuel.

One important application of cutting is in underwater operations such as demolition work, marine salvage work or structural repair work. Since acetylene is suitable only at depths down to 30 feet, hydrogen is normally used as the fuel gas. A special design of nozzle is used which provides a shroud to protect the flame from the water.

The oxygen cutting process is excellent for mild steel and is very widely used either manually or mechanically. With alloy steels, however, it is often impossible to produce a reasonably cut surface because the alloying ingredients form refractory oxides which hinder contact between the oxygen and the iron. In cutting such alloys it is necessary to introduce either iron powder or a flux into the cutting stream. The refractory oxides are removed by a combined fluxing and melting operation.

There is also difficulty in cutting cast iron owing to the relatively high proportion (about 3 per cent) of graphitic carbon which it contains. Carbon dioxide is produced in sufficient quantities from this carbon to dilute the cutting oxygen and so reduce the rate of oxidation of the iron. In addition the melting point of cast iron is about 1200°C against about 1550°C for mild steel. The melting point of FeO is 1419°C and that of Fe_3O_4 is 1538°C; consequently the surface of cast iron is not automatically cleaned of the oxide.

Of other metals only titanium has an oxide of lower melting point than the metal, so that apart from this metal only rough severing can be achieved by the use of oxygen alone.

It should be mentioned that there are several other operations performed on mild steel which depend on the same principles as those of cutting. These include gouging in which imperfections such as faulty welds or sand inclusions are removed; scarfing or deseaming in which surface defects on bars, billets, blooms, etc., are removed; surface planing; and hot cropping which replaces shearing and sawing. These applications require torches of special design.

OTHER OXY-FLAME PROCESSES

Several other processes use the high temperatures produced by oxygen flames. For example, flame hardening of steel involves passing the article to be hardened close to a suitable flame and then rapidly cooling by a water spray. The result is a hardening of the surface only of the metal, so enabling certain items such as gear-wheels to have a much longer life.

Flame cleaning is a method of cleaning iron and steel surfaces prior to painting. A thin but wide oxy-acetylene flame or group of flames is moved along the surface of the metal. Differential expansion takes place between the rust and any remnants of old protective layers on the one hand and the parent metal on the other, causing the former to become detached. The debris is partly blown away and partly removed by a wire brush, leaving a clean warm surface particularly suitable for immediate painting.

MANUFACTURE OF STEEL

Before World War I the major use of oxygen was in welding operations. Between the wars cutting operations developed enormously while competitive welding processes such as arc welding were introduced. Since World War II and especially since 1955, the largest single use of oxygen has grown to be in the manufacture of steel.

The two primary steel-making processes, the open hearth and the Bessemer or converter processes, were developed in the period 1860–80. In the United Kingdom most steel was made by the open-hearth process and a smaller amount was made in side-blown Bessemer (or Tropenas) converters. On the Continent most steel was made in bottom-blown Bessemer converters because of the necessity of using high phosphorus irons, although this steel is inferior to open-hearth steel on account of its lower carbon and higher nitrogen content.

The availability of oxygen at low cost and in large quantities has radically altered the methods used by the steel industry. A selection of the ways in which oxygen is now used is as follows:†

Open-hearth furnace. By enriching the combustion air with oxygen more fuel can be burnt in a furnace of a given volume without increasing the

† See references 7 and 8 in the Bibliography for a more detailed account.

waste gas volume. Consequently higher melting rates can be achieved, which is of particular importance in plants using a considerable quantity of steel scrap as raw material as well as pig iron.

The chemical reactions occurring in the open-hearth furnace involve the successive oxidation of silicon, carbon and phosphorus as well as of sulphur and manganese. If the molten blast-furnace metal is treated with oxygen while it is in the ladle transferring it to the open-hearth furnace, a considerable amount of the silicon can be oxidized so reducing the load on the furnace. In the furnace itself, the use of oxygen instead of air for oxidizing the carbon also reduces the refining time. It should also be noted that, because of the exothermic nature of this reaction there is some reduction in the amount of fuel required to heat the furnace.

The tap-to-tap cycle time of the furnace is considerably reduced by the use of oxygen, the actual reduction depending on the procedure employed. Typical figures are a reduction from 12 hours to 6 in one case and 9 hours to 4 in another.

Converter process. In this process little or no external heat is employed, so that there is no opportunity for increasing fuel efficiency by the use of oxygen. The Thomas–Gilchrist adaptation of the Bessemer process uses a basic furnace lining and has chiefly been used with high phosphorus iron; it relies on the oxidation of the phosphorus for thermal support. Because of the additional heat required for melting it has not been possible in normal practice to use much scrap with the result that such converters were rather inflexible as regards their raw materials. Moreover, because of the nitrogen present in the air used for blowing, the steel tended to contain about 0·015 per cent of nitrogen which is more than is desirable for many purposes. The use of oxygen, or even oxygen-enriched air, improves the thermal output and enables considerable proportions of scrap to be used as well as providing a steel of lower nitrogen content.

Linz–Donawitz process. This process was developed in Austria after World War II for operation on irons containing up to 0·3 per cent phosphorus. It consists of a converter in which oxygen, instead of being blown into the metal from below as in other processes, is jetted at high velocity at the metal surface from above. A high surface temperature is produced and, as in the Bessemer process, phosphorus oxidation starts before elimination of carbon has been completed. Considerable proportions of scrap can be used and the product is low in nitrogen.

Rotor process. This process was developed in Germany. A horizontal, cylindrical furnace rotates about twice per minute. Oxygen is introduced both above and below the metal surface.

Kaldo process. This process was developed in Sweden. The furnace is inclined at an angle of about 15° to the vertical and is rotated about thirty times each minute. Because of the good contact between slag and metal early oxidation of phosphorus occurs.

Electric furnaces. In the United Kingdom this type of furnace is primarily used to make high-grade and alloy steels. The amount of power used to melt the charge can be greatly reduced by blowing in oxygen. In addition the melting period is shortened so enabling a greater annual output to be achieved in the furnace. The faster removal of carbon also reduces the turn-round time.

MANUFACTURE OF IRON

The blast furnace of standard pattern is a highly efficient piece of equipment for making iron from its ores. The heat content of the rising gases is transferred to the descending solid burden, but in order to achieve high efficiency the height of solid must be considerable. This in turn means that the principal solid component, coke, must be sufficiently robust to resist crushing. Not all iron-making countries have coals of a quality suitable for making strong metallurgical coke.

If oxygen-enriched air is used in the blast furnace less heat is transferred to nitrogen and so adequate heat can be transferred to the solid burden in less height. This enables a weaker coke and also a poorer-quality ore to be used. In addition the gas leaving the furnace has a higher calorific value because of the elimination of some of the nitrogen. A number of experimental low shaft blast furnaces have been successfully operated and particularly one at Ougree near Liège, Belgium, under the auspices of the Organization for European Economic Co-operation. So far, however, the use of oxygen has not been adopted on a large scale.

GASIFICATION OF PRIMARY FUEL

If a basic solid fuel, such as coke, is heated with steam and air it can be gasified without external heat being supplied, by the two reactions:

$$2C+O_2 \rightarrow 2CO \quad \text{exothermic}$$
$$C+H_2O \rightarrow CO+H_2 \quad \text{endothermic.}$$

In order to avoid contamination of the product by nitrogen the process is operated on a cycle lasting 3 to 6 minutes depending on the design of plant. During the first stage air is blown through the coke raising the temperature, and during the second stage steam is blown through the hot solid yielding a gas containing approximately equal proportions of carbon monoxide and hydrogen. This process, known as the water gas process, has been used for many years.

If, instead of air, oxygen is used there is no need to operate a cyclic process and a more efficient and speedy process results. Further, if the process is operated under a pressure of 20 atm (Lurgi process) then some of the primary products are converted to methane by the reaction

$$2CO+2H_2 \rightarrow CH_4+CO_2.$$

In practice it is possible to make a gas containing about 20 per cent methane and closely resembling town's gas in its chemical and physical properties. Although this process has been used in certain other countries, only two works have operated the process in the United Kingdom. Indeed in view of the discoveries of natural gas under the North Sea interest in the process has faded.

Alternative feedstocks which can be used are petroleum fractions and natural gas or methane, usually at atmospheric pressure. It should be pointed out that once a carbon dioxide/hydrogen mixture has been produced its composition can be adjusted by using the "water gas shift" reaction:

$$H_2+CO_2 \xrightarrow{\longleftarrow} CO+H_2O.$$

Such mixtures are needed for processes such as the production of methanol or the Fischer–Tropsch process. The reaction can also be used to make the hydrogen needed for synthetic ammonia manufacture.

OXYGEN FOR BREATHING

Although the amount of oxygen used for this purpose is a very small part of the total manufactured (about 0·02 per cent in the United Kingdom) it is of vital importance to human life. If pure oxygen is breathed instead of air there is much less strain on the heart. It is therefore used in hospitals in two main ways. Firstly, it is used in oxygen tents and in incubators for premature babies. Secondly, it is used in conjunction with anaesthetic gases during operations.

While cylinders are still widely used for hospital supplies, some of the larger hospitals have oxygen piped to appropriate points from a central supply which may either be batteries of cylinders or liquid oxygen evaporators.

Oxygen for breathing is also supplied to high-flying aircraft for either military or civilian use. While compressed oxygen carried in cylinders can be used, it is now usual for liquid oxygen in small containers to be carried on the aircraft because of the saving in container weight. The system is so arranged that as soon as the pilot or passenger starts to breathe at his mask, liquid oxygen starts to evaporate and the evaporated gas is warmed to room temperature.

Oxygen for breathing is of course essential in spacecraft.

LIQUID OXYGEN EXPLOSIVES

If a cartridge of a cellulosic material such as sawdust is soaked in liquid oxygen immediately prior to use, it can be loaded and fired in the same way as conventional solid explosives. Such explosives can have a performance better than 60 per cent gelignite, and have the technical advantages that they are perfectly safe to handle except between the soaking vessel and the firing hole, and that if they should fail to fire the operators have only to wait a limited time before the oxygen has evaporated and the cartridge is again quite safe to handle. Naturally thefts of such cartridges do not occur.

The economics are most favourable where regular blasting takes place and the liquid oxygen demand can be forecast with accuracy. The liquid is stored in a site tank from which it is fed under gravity into transportable soaking vessels.

Its use is currently confined to opencast mining in India and underground iron-ore mining in France. In both these cases it is used on a considerable scale.

ROCKETRY

All large liquid fuelled rockets use oxygen as the oxidant whether the fuel is liquid hydrogen or kerosene. The liquid oxygen/hydrocarbon system has the advantage for the first stage of large propulsion units that the materials are cheap and readily available. For the second and later stages the liquid oxygen/liquid hydrogen system has the highest specific impulse (thrust per mass rate of fuel consumption) of any known system apart from exotic combinations such as hydrogen and ozone or hydrogen and fluorine. The main disadvantage is the low density of liquid hydrogen. Care has to be taken because of its flammability but there are no toxic hazards. The process of liquefaction of hydrogen is described in Chapter 12, p. 92.

CHAPTER 7

Nitrogen

METHODS OF PRODUCTION

Nitrogen is produced commercially in three ways, namely by cracking of ammonia, by inert gas generators and by distillation of air. The choice of method depends on the quantity and purity of product required.

When only modest amounts, say 50–250 ft³ per hour, are needed ammonia dissociation may be employed. The ammonia, usually stored as liquid under pressure, is passed at 500–600°C over a catalyst. The hydrogen formed in the process can be partly or wholly removed by burning in air. The two operations may be carried out simultaneously using, for example, a platinized alumina catalyst together with a rhodium catalyst supported on alumina. The water formed in the process is removed by condensation. Since some oxides of nitrogen are liable to be formed it may be necessary to remove these in the case of certain uses of the nitrogen gas. Unless considerable trouble is taken in purification the final product always contains some hydrogen. This is quite acceptable for certain purposes and in certain metallurgical operations is even desirable because it helps to avoid oxidation.

Inert gas generators are made in the range 250–50,000 ft³ per hour. They depend on the combustion of a fuel with the exact amount of air needed, in combustion chambers which are horizontal, firebrick lined and water jacketed. The commonest fuels are town's gas, gas oil and diesel oil, although propane, butane, hydrogen, producer gas or water gas are sometimes used. Two typical compositions of the gas produced using town's gas and oil are given in Table 5.

TABLE 5. NITROGEN PRODUCT FROM INERT GAS GENERATORS

	Town's gas, per cent by volume	Oil, per cent by volume
Nitrogen	87	84
Carbon dioxide	12	15
Oxygen	1	1
Carbon monoxide	1	1
Oxides of nitrogen	0·0075	0·015

In addition since the fuels contain some sulphur the product contains some oxides of sulphur. In some generators a mixture of steam and air is used with the result that the product contains some hydrogen.

It will be seen that inert gas generators produce a rather impure product in the first instance, although it can be used in certain operations as an inert atmosphere. It is, however, common practice to remove some of the impurities and depending on the purity of product required the methods of purification can take a number of forms. Thus counter-current scrubbing with ethanolamine in a packed tower removes carbon dioxide and other acidic gases. Any non-acidic oxides of nitrogen, hydrogen or carbon monoxide must be oxidized prior to removal. While the principal costs in producing crude inert gas are fuel, power and capital charges, the cost of subsequent purification can be considerable, depending on the purity required.

When a high purity gas is required or when the scale of operation is considerable, then the low-temperature air separation method is employed. Most nitrogen made commercially, other than for synthetic ammonia, is made by the low-temperature route and has a purity of at least 99·5 per cent. The small consumer gets the advantage of bulk production and distribution. Where liquid nitrogen is required the low-temperature route is always used.

In the liquid-oxygen plant described in Chapter 6 it was mentioned that the waste nitrogen fraction leaving the plant contained 2–3 per cent of oxygen. The major reason for this is that argon boils between oxygen and nitrogen, and must leave the column either with the oxygen or the nitrogen. If, however, the plant is required to produce only nitrogen then a plant similar to the oxygen plant can be used and the column operated to produce pure nitrogen at the top and oxygen of 90–95 per cent purity at the base. More usually, however, the plant is required to produce oxygen together with a smaller amount of nitrogen. While it is impossible to produce *all* the nitrogen from the oxygen plant in pure form it is possible to produce *some* pure gaseous nitrogen (say up to 50 per cent). This is done by introducing more plates into both top and bottom columns, and arranging to withdraw pure nitrogen at the top of the upper column and impure waste nitrogen at a point rather lower down as shown in Fig. 7.1. An alternative method is to use the standard

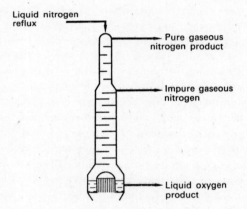

Fig. 7.1.   Pure gaseous nitrogen attachment.

oxygen column and pass all the impure nitrogen to a column of smaller diameter in which some pure nitrogen can be separated.

Some customers require liquid nitrogen and if the amounts needed are small they can be withdrawn from the pure liquid nitrogen stream as it leaves the lower column (see Fig. 4.5). This has the effect of reducing the amount of liquid available for reflux in the upper column, and if not strictly limited would have an adverse effect on oxygen production. When larger amounts of liquid nitrogen are needed an additional fractionating column is introduced into which the impure nitrogen leaving the main column passes (Fig. 7.2). Liquid oxygen at a lower pressure is used to condense the nitrogen at the top of the subsidiary column. Since the oxygen is vaporized in the process the ratio of oxygen gas to liquid leaving the plant is increased.

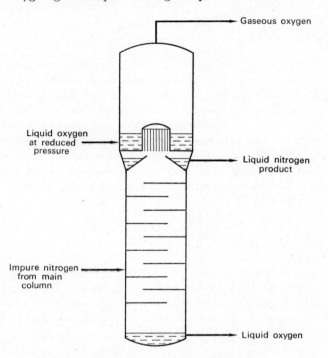

Fig. 7.2. Production of liquid nitrogen.

The British Standard Specification for Industrial Nitrogen† defines two grades. For the general grade the oxygen content should not exceed 0·5 per cent, the carbon dioxide plus other carbon compounds calculated as carbon dioxide 10 ppm, and hydrogen 20 ppm by volume. The special grade requires the oxygen not to exceed 10 ppm, the carbon dioxide plus other carbon compounds calculated as carbon dioxide

† See reference 9 in the Bibliography.

10 ppm and hydrogen 10 ppm by volume. The specification also requires the water vapour content not to exceed 60 ppm initially. In the case of cylinder nitrogen the water vapour content will rise as the cylinder empties but should not reach more than about 600 ppm finally.

USES OF NITROGEN

Nitrogen is used industrially in both gaseous and liquid forms. Lavoisier, who first established the composition of air, gave nitrogen the name "azote" meaning lifeless. It is indeed relatively inert chemically at room temperature and is widely used as an inert atmosphere. However, at higher temperatures it does take part in a number of important chemical reactions. While the uses of gaseous nitrogen fall into one of these two categories, the uses of liquid nitrogen, although smaller, are more varied. The United Kingdom consumption of nitrogen, other than for tied uses, such as the production of fertilizers, was about 615,000 tons in 1972.

CHEMICAL USES

The natural fixation of nitrogen through the nitrogen cycle is essential for life. This cycle includes the assimilation of elementary nitrogen by some micro-organisms or the combination of oxygen with nitrogen by electric discharge, followed by the washing out of nitric acid by rain. In the last 75 years the artificial fixation of nitrogen has been practised industrially. The most important method is the reaction of nitrogen with hydrogen to give synthetic ammonia as described on p. 91. It should be noted that this is only one way of achieving fixation and that a slightly earlier method, patented by Frank and Caro, in 1895, involves the reaction of nitrogen with calcium carbide at atmospheric pressure and a temperature of about 1100°C to give calcium cyanamide:

$$CaC_2 + N_2 \rightarrow CaCN_2 + C.$$

This material was originally used as a fertilizer and for conversion to ammonia, urea and black cyanide. While it is not manufactured in the United Kingdom, it is still made in certain other countries where it is either used as a fertilizer or hydrolysed to dicyandiamide for conversion to melamine, an important material in the plastics industry. The reactions are :

$$2CaCN_2 + 4H_2O \rightarrow 2Ca(OH)_2 + C_2H_4N_4 \text{ (dicyandiamide)}$$

$$3C_2H_4N_4 \rightarrow 2C_3H_6N_6 \qquad \text{(melamine)}$$

Guanidine and thiourea are also made from cyanamide.

Nitrogen can also be used to make metallic nitrides some of which, e.g. those of titanium, zirconium and tantalum, are of industrial importance because of their hardness and resistance to corrosion. Others such as those of boron, silicon, aluminium and gadolinium are ceramic-like, non-scaling and resistant to heat and corrosion as well as being semiconductors. The first group mentioned can be made by reacting nitrogen or ammonia with the appropriate metallic oxide in the presence of carbon. Others, such as those of silicon, aluminium and magnesium, can be made by reacting nitrogen or ammonia with the metal or metal hydride.

INERT ATMOSPHERES

The non-chemical uses of nitrogen gas depend mainly on it replacing air so as to prevent oxidation occurring. The largest single use of this type is in the bright annealing of stainless steel. The nitrogen may be used alone or in admixture with hydrogen which acts as a deoxidizing agent; a typical mixture may contain 3 per cent of hydrogen. Bright annealing is also used by steel mills for tin plate and by the electronics industry in the manufacture of electron tubes and the preparation of glass to metal seals.

There is a growing use of nitrogen gas in packaging, particularly of foodstuffs. This prevents loss from bacterial action, or even fire, and also preserves flavour and aroma. Large numbers of materials are handled in this way including milk, milk products, fruit, fruit juices, potato chips, dried coconuts, beer, wines, powdered eggs, dehydrated vegetables, edible fats, nuts and nut products, meat products, biscuits, fish products and coffee.

The chemical industry makes use of nitrogen atmospheres in a number of cases involving flammable materials and to prevent undesirable side reactions. The blanketing of petrol during storage reduces gum formation. In polypropylene manufacture traces of moisture can be harmful and about 200 ft³ of nitrogen are said to be used per ton of product produced. In nylon manufacture nitrogen is used at the rate of 1000–2000 ft³ per pound of polymer for blanketing the nylon salt fed to the reactors.

Nitrogen blanketing is also used to prevent oxidation in certain of the more sophisticated metallurgical processes. Thus in the manufacture of germanium transistors it is used both as an inert atmosphere and as a carrier for "doping", i.e. the addition of controlled amounts of impurity to the germanium. Nitrogen can only be used for this sort of application providing no reaction occurs with the metal. If it does then the more expensive argon must be used instead.

Nitrogen gas finds use in transmission cables in the electrical distribution industry. When voltages exceed 66,000 a solid-type cable is unsatisfactory both technically and economically because ionization takes place in voids or gas pockets inside the dielectric. Cables operating at these high voltages are filled either with oil or nitrogen, normally at a pressure of about 14 atm. The use of nitrogen enables the dielectric wall thickness to be reduced resulting in a smaller and lighter cable. It also reduces thermal resistance so that an increased current loading can be used. Heat transfer from the conductor is also improved thus preventing an undue temperature rise.

Still another application of nitrogen blanketing is in the process for making float glass recently developed by Pilkington's. In this the glass is floated on molten tin which is protected from oxidation by an atmosphere of nitrogen. It has been stated that 2800 ft³ of nitrogen is used per ton of glass.

USES OF LIQUID NITROGEN

The number of uses of liquid nitrogen has grown rapidly in recent years and so has the consumption of liquid nitrogen. The applications depend on the refrigerant properties of liquid nitrogen and its ready availability in the more highly industrialized countries. Since liquid nitrogen boils at $-196°C$ it is inherently more expensive than means of refrigeration which do not achieve such low temperatures. If only a modest level of refrigeration—say a few degrees below the freezing point of water—is required, then a normal refrigerating machine or even solid carbon dioxide is likely to be cheaper. However, one special case in which it is used, although only a modest level of refrigeration is needed, is in food distribution, on account of its advantage over other methods in that it can be used for very rapid cooling.

Food Distribution

If a delivery van is required to be maintained at a temperature of, say, $-2°C$ while transporting frozen food and delivering to a range of customers, a conventional refrigerator unit operating off the engine of the vehicle would take several hours to cool the van down initially. Moreover, every time the van doors were opened the atmosphere in the van would warm up, particularly on a hot summer's day, and the contents would begin to warm up. Again the conventional refrigerator would take a long time to bring the temperature back to the required figure.

When a low-temperature refrigerant like liquid nitrogen is used only a small amount has to be introduced and evaporated to cool the van down rapidly and to maintain an even temperature despite opening of doors, because of its high "cold" content. Arrangements are made for the temperature to be automatically controlled by linking a temperature measuring device to a liquid nitrogen inlet valve situated at the top of the refrigeration space. This acts rather like a fire sprinkler and releases a stream of liquid nitrogen droplets and cold gas which fall, so cooling the space rapidly.

This process is sometimes referred to as Polarstream and in the United States it has also been used on rail freight cars. Not only is less precooling necessary and a more even temperature maintained but the mechanical equipment required is simpler and less moisture is lost by evaporation than occurs in the forced convection inherent in the mechanical refrigerator.

Living Matter

It is frequently found that rapid cooling of living cells to liquid nitrogen temperature results in a condition of suspended animation in that when the material is warmed up to room temperature again, even after a considerable time has elapsed, it is completely unimpaired.

A common example of suspended animation in connection with food supplies relates to artificial insemination. This procedure has, of course, been practised for a long time with horses and dogs and especially cattle. At atmospheric temperature and even at temperatures a little below the freezing point of water the semen has a limited life. If, however, fresh bull semen is immediately cooled to liquid nitrogen temperature it can preserved for many years—perhaps a hundred or more. In this way selected bulls can father offspring for many years. There are about fifty semen banks in the United Kingdom and very many more in the United States, where the preservation of bull semen is almost entirely a cryogenic operation.

Liquid nitrogen can be used to preserve human blood almost indefinitely as against the normal period of about 3 weeks. It is essential for it to be cooled to liquid nitrogen temperature in a minute or less, and also to be heated up very rapidly to blood temperature when required for use.

Liquid nitrogen has also been used by the medical profession in a number of other ways. It can be used to preserve bone marrow, glands and tissues. It can be used to lower the body temperature to such an extent that the heart has less work to do in circulating oxygen, so helping in serious heart operations. It can be used in brain operations by applying liquid nitrogen with pinpoint accuracy to a small part of the brain which can then undergo an operation in which no blood flows because it is frozen. Another example is the destruction of a cancerous growth by alternative freezing and thawing. These various surgical applications are sufficiently widespread for the term cryosurgery to have come into common use.

Mechanical Operations

The properties of many substances are changed at liquid-nitrogen temperatures in a direction which enables practical use to be made of the change. For example, all metals contract on cooling and this enables a process known as shrink fitting to be performed. If a metal part has to be inserted tightly into a hole, this can be achieved by forcing it in by some mechanical means if the scale is not too large. On a large scale, e.g. insertion of a liner into a large gun barrel, or the shaft into a giant drum for use in drag-line operations in open-cast coal mining, the objective can be more readily achieved by cooling the part to be inserted by immersing it in liquid nitrogen. The cold part is then placed carefully in position and allowed to warm up, by which time it will be rigidly in position.

When underground pipes have to be repaired a trench is dug to enable the workers to get at the pipe. Certain soils, particularly in wet weather, tend to slip and fill the hole, so impeding the repair work. Treatment with liquid nitrogen will freeze the water in the soil and the trench will remain clear for several hours, which is generally sufficient time to complete the work.

Synthetic plastic materials and also rubber become hard and brittle at liquid-nitrogen temperature. They can then be ground to give a powder which can later be moulded into desired shapes. Moulded rubber articles often have flashes, or unwanted pieces of rubber, attached to them. Instead of the normal practice of removing the flashes by a cutting operation, the articles can be cooled in liquid nitrogen and then tumbled in a cylinder. The impact of one frozen article on another causes the flashes to drop off.

Metallurgy

The properties of some metals change at liquid nitrogen temperature and workers in the field of metal fabrication have found that desirable properties can be introduced into metals in some instances. Not all these are being used industrially at the moment but a brief account of some of the possibilities is given below.

The machining of certain special alloys generates a large amount of heat at the interface of the tool and the metal being worked. Apart from local work hardening and oxidation of the metal, the tool life is lowered.

Cooling by carbon dioxide is sometimes practised but cooling by liquid nitrogen, although more expensive, is convenient and at a lower temperature level. Tool life can be greatly increased and the surface effects on the metal diminished.

If certain stainless steels are rolled at liquid-nitrogen temperature instead of at room temperature, there is an increase in tensile strength, yield strength and hardness. This phenomenon is sometimes referred to as "zero-rolling". If, after the cooling treatment, the steels are heated to 425°C additional strengthening takes place.

Another process, known as "cold stretching", is carried out on fabricated vessels. All openings to the vessel are sealed except for a single gas connection. The vessel is then cooled in liquid nitrogen, and gaseous nitrogen under pressure introduced through the gas connection. The pressure can be maintained until the vessel has stretched by about 15 per cent into a die of the desired final shape and size. On warming to room temperature the vessel is found to have increased its tensile strength threefold. This high strength-to-weight ratio helps steel to compete with filament-wound glass fibre vessels in applications where this factor is important, for example in rocket casings, torpedoes or special gas-storage spheres.

Aluminium alloys can be stress relieved at low temperatures. The materials so treated are less likely to give trouble in subsequent operations such as machining.

CHAPTER 8

Argon

THE discovery of the inert gases has been mentioned in Chapter 2. Ramsey obtained argon by passing air in turn over soda lime to remove carbon dioxide; phosphorus pentoxide to remove water vapour; hot copper to remove oxygen; and hot magnesium to remove nitrogen. Nowadays it is produced commercially by the low-temperature distillation of air. The inert gases were at one time described as "rare" gases. Argon is certainly not a rare gas today. Consumption in the United Kingdom in 1971 was of the order of 18,000 tons or 380 million ft³, with a growth rate of about 16 per cent per annum.

PRODUCTION

Argon boils at a temperature intermediate to oxygen and nitrogen, and in an air distillation column tends to accumulate at a point rather below the feed point to the upper column (Fig. 4.5). The degree of accumulation depends on a number of factors such as the purity of the oxygen and nitrogen leaving the column. At the point of highest argon concentration the vapour phase may contain 15–20 per cent, the balance being oxygen. If a side stream is withdrawn from this point (Fig. 8.1) it can be passed to a subsidiary small column for further distillation, giving a product containing 80–98 per cent argon. The residual oxygen can then be removed by reaction with hydrogen, followed by removal of the water vapour produced. A further fractional distillation column enables excess hydrogen and any nitrogen to be removed. The refrigeration for this last stage is provided by liquid nitrogen from the plant. Indeed the product can be withdrawn as pure liquid argon, if required

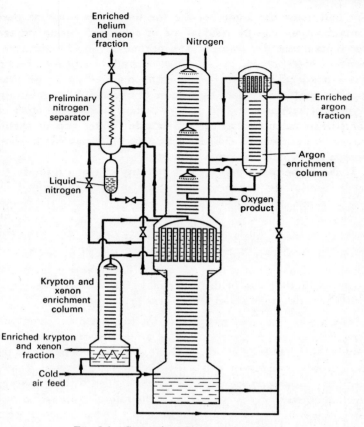

FIG. 8.1. Separation of inert-gas fractions.

in this form for transport. This method of isolation enables yields of argon of about 50 per cent of that entering the plant to be obtained. Higher yields are only possible with more complex, and therefore more costly, plant. This is not justified while argon demand is only a small fraction of the argon potentially available as a by-product of oxygen production.

Since the uses of argon are mainly in operations which can tolerate neither oxygen nor nitrogen, purity is of fundamental importance.

Not only must the argon leaving the separation plant be pure but contamination, e.g. by residual air in cylinders, during storage and distribution must be avoided. The normal argon of commerce has a purity of 99·99 per cent, but an industrial grade of 99·995 per cent purity is also made. The limits of impurities required by British Standard specification† are 10 ppm by volume of oxygen, 50 ppm of nitrogen, 10 ppm of hydrogen and 10 ppm of carbon dioxide and other carbon compounds calculated as carbon dioxide. Water vapour should not exceed 30 ppm when a cylinder is first used but may rise to about 600 ppm as the cylinder becomes exhausted.

It should be noted that a small quantity of argon is produced as a by-product of the commercial synthesis of ammonia from hydrogen and nitrogen. In this process crude nitrogen from air is reacted with hydrogen over a catalyst. If the gases are recycled in a closed system the unreactive argon, present as an impurity in the nitrogen, tends to accumulate and can be collected for purification and use.

USES

Argon was first used industrially for filling electric-light bulbs. A typical incandescent lamp consists of a filament of tungsten heated to 2800°C in the absence of oxygen. If an evacuated bulb is used the filament slowly evaporates and is deposited on the inside of the glass envelope, causing darkening. If the bulb is filled with an inert gas the thermal conductivity is reduced as is the rate of evaporation so that a lamp of much longer life and operating at a higher temperature can be produced. The first gas-filled lamps used nitrogen but it was later found that argon, because of its higher molecular weight, reduced evaporation and heat losses still further. Because with pure argon some arcing may occur it is customary to use a mixture of 88 per cent argon and 12 per cent nitrogen. The pressure in the bulb when cold is 0·8 atm, so that in use the heating of the gas by the filament raises the pressure to about atmospheric.

Nowadays in the United Kingdom the largest single use of argon is in

† See reference 11 in the Bibliography.

welding processes, employing either a consumable or a non-consumable electrode. In the latter process an electrode such as tungsten is surrounded by an envelope of argon gas, the arc being struck between the electrode and the metal to be welded as shown in Fig. 8.2. A rod of

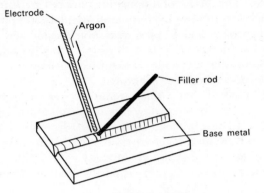

Fig. 8.2. Argon arc welding.

filler material is used if necessary. The heat of the arc melts the base metal and filler rod producing a uniform weld. The argon protects the hot metal from the action of the air. This process has been used particularly in the welding of aluminium and its alloys.

In the second process a consumable electrode is used instead of the tungsten one. It consists of a coiled wire of similar composition to the base metal fed continuously and automatically at a controlled rate. The electrode and base metal melt together protected by argon.†

In addition to the above applications of argon there are a variety of uses in the metallurgical and allied industries wherever an atmosphere free from both oxygen and nitrogen is needed. These include the manufacture of reactive metals such as titanium and zirconium; manufacture

---

† The reader should note that all welding processes provide some protection of the base metal from the action of air. In oxy-acetylene welding shielding is mostly provided by the gaseous combustion products. In welding with stick electrodes the weld is covered and protected by the melted electrode covering, or in some cases by the gases produced. In submerged arc welding protection is afforded by the flux which floats on the weld puddle and is later chipped off.

of transistor materials such as germanium and silicon; and melting and casting of special materials.

It should be mentioned that for all these applications, except lamps, helium can be used. This is rarely done in the United Kingdom where the price of helium is about six times that of argon. It is sometimes done in the United States where helium was freely available earlier than argon and the price differential is quite different.

CHAPTER 9

Helium

HISTORICAL

Helium is a unique element in a number of ways which will be mentioned later in this chapter. It is also unique in being known to exist nearly 30 years before it was isolated. Line spectra, i.e. spectroscopic lines characteristic of the elements in a hot body, were first observed in 1860. In 1868 during a total eclipse of the sun Hansen observed lines which had not been observed elsewhere. These lines were attributed by Lockyer to an unknown element which he called helium. The discovery of the inert gases by Ramsey in the period 1893–5 has been referred to in Chapter 2. One of these inert gases was found to have the same line spectra as the helium observed in the sun. At about the same time it was found that certain minerals such as clevite (Norway) and uraninite (U.S.A.) contained occluded gases rich in helium. It was then found that certain natural gases in Canada and the United States contained helium, sometimes in quantities as high as 1–6 per cent. These are 10,000 to 60,000 ppm compared with 5 ppm in air. Such gases have been used in the bulk production of helium.

The first plant erected to isolate helium was at Hamilton, Ontario, in 1917 in order to supply Britain with the gas for dirigibles and balloons. Because of the poor raw material available in Hamilton the plant was moved to Bow Island, near Calgary, in 1918 but was shut down in 1920. Although the United States started to isolate helium somewhat later they were in a much better position both as regards the quality and the quantities of helium-rich gases. They expanded their production steadily and were the sole suppliers of helium in the Western world until 1963 when a plant was again built in Canada to exploit finds of helium-rich gas at Swift Current, Saskatchewan.

The United States Government operates a helium-conservation programme. Indeed for many years they were themselves the sole producers of helium in the United States. Most of the helium consumed in that country is for government purposes, for modern technology and research. Through legislation control is exercised over helium exports and the production of helium is encouraged, particulary from natural gas being piped for combustion, as a result of which the helium would be lost to the atmosphere. Most natural gases do not contain enough helium to make extraction worth while. Those that do vary considerably in composition as will be seen from two typical analyses of gases from which helium is extracted:

	Helium	Nitrogen	Methane	Higher hydrocarbons	Carbon dioxide
Gas A	1·4	12·7	78·2	7·7	—
Gas B	2·1	26·3	64·2	6·6	0·8

PRODUCTION

Since helium boils at a temperature considerably below all the other components of natural gas, the method of isolation is to cool the gas to a temperature at which all the other substances will condense, leaving the helium in the gas phase. The process of separation is thus fundamentally simpler than that of air separation in that no distillation stage is needed.

The stages involved are shown diagrammatically in Fig. 9.1. The gas is first compressed and easily condensable materials present, such as water vapour, carbon dioxide or hydrogen sulphide, are condensed or washed out with a solvent, e.g. monoethanolamine. Passage through activated alumina removes traces of water vapour. The purified gas is then cooled by heat exchange with previously liquefied gas leaving the plant. Further cooling is obtained partly by expanding the natural gas

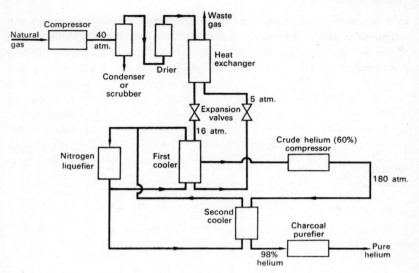

Fig. 9.1. Helium separation from natural gas.

itself and partly by a supplementary nitrogen cycle. Complete separation is not possible in a single stage, and in practice three stages are used, intermediate products containing 60 and 98 per cent helium. The normal grades sold in the United Kingdom have purities of 99·95 and 99·995 per cent.

A little helium is still made in countries other than the United States and Canada as a by-product of the separation of neon from air (see Chapter 10).

The present scale of helium production in the United States is 1016 to 1082 million ft³ per annum and consumption is in the range 867 to 1000 million ft³. The consumption in the United Kingdom has risen rapidly in recent years, e.g. from 0·2 million ft³ in 1962 to 12·5 million ft³ in 1968. It is currently imported from North America either as gas in cylinders by sea, or as liquid by sea or air.

FORMS OF HELIUM

Helium has two isotopes, the more abundant having an atomic weight of 4 and the less abundant one of 3. The respective boiling points are −268·93°C and −269·98°C. It is of interest that the abundance ratio of the two isotopes, i.e. the ratio of H³ to H⁴, varies with the source of the helium :

Earth's atmosphere	12·0	10^{-7}
Twelve natural gas wells	0·5–3·0	10^{-7}
Five radioactive minerals	0·3	10^{-7}

Research workers requiring He³ may therefore prefer to start with atmospheric helium if it is obtainable. It should be noted that H³ is also obtainable from certain radioactive processes.

A number of properties of liquid helium change markedly below −271°C, referred to as the lambda point, and it behaves as if it were a different substance. Helium below this temperature is called helium II.

USES OF HELIUM GAS

As mentioned above the first helium-production plant was built to provide a light incombustible gas to fill balloons and dirigibles. Although helium has twice the density of hydrogen it still has a density only one-seventh that of air and it is also perfectly safe. Dirigibles are no longer used but helium is still used in balloons sent up for various purposes such as weather observations and cosmic-ray studies.

The use of helium instead of argon in gas-shielded arc welding and in the protection of certain non-ferrous metals during processing has been referred to in Chapter 8. However, it cannot compete economically except in certain instances in the United States and Canada.

When deep-sea divers are under water they breathe gas pressurized to the pressure to which they are subjected at any particular depth. Under these conditions nitrogen dissolves in the blood to an appreciable extent. When the diver is brought to the surface the nitrogen comes out of solution and causes intense pains behind joints such as knees and elbows known as "diver's bends". Helium is much less soluble in blood than nitrogen and a mixture of 80 per cent helium and 20 per cent oxygen is supplied as breathing gas for divers.

Another interesting use of helium is to pressurize the tyres of large planes, in place of air. This so reduces the weight of gas that one extra passenger can be carried.

LIQUID HELIUM

The most interesting and unusual properties of helium are at low temperatures in the liquid form. Potentially there are a number of commercial applications, although use has so far been confined to a few examples. There are several thousand scientists with their assistants in various countries carrying out research involving liquid helium, so that it can be expected that as a by-product of this work some practical uses will ultimately emerge on a substantial scale. Some account of the possibilities will be given in the next few pages.

The liquefaction of helium is more difficult than that of any other gas. If use is to be made of the Joule–Thomson effect the gas has first to be cooled to below the inversion temperature of −233°C. The only practical way of doing this is by a bath of liquid hydrogen. Hydrogen in its turn has an inversion temperature of −69°C which involves cooling the hydrogen to below this temperature in a bath of, say, liquid nitrogen. For these reasons the use of expansion machines for performing external work is favoured in commercial liquefiers. Although no preliminary cooling is required there are a number of mechanical problems in using expansion machines at these low temperatures.†

Methods adopted for storing liquid helium have been described in Chapter 5.

SUPER CONDUCTIVITY

As the temperature of any metal is lowered its electrical resistance decreases which means that its conductivity increases. In the case of some metals and alloys the resistance drops to nothing a few degrees above absolute zero and the metal is said to be superconducting. The actual temperature at which this happens varies from metal to metal but is in the region below −255°C. This phenomenon may become of

---

† For further information on this subject see reference 4 in the Bibliography.

immense importance in a number of fields providing certain practical problems can be overcome.

The first application was made during World War II in a bolometer for infra-red detection. A bolometer is a sensitive temperature-measuring device which operates by absorbing radiant energy in a metal whose resistance is then measured. If the temperature of the metal can be controlled at that at which superconductivity is just beginning to occur, the resistance change per unit of heat absorbed is very high and an extremely sensitive instrument can be produced. This device, using a niobium nitride strip maintained at $-258°C$, was used to detect the presence of enemy troops and equipment.

The possibility of operating an electromagnet in a superconducting state, so that the current would circulate endlessly and the magnetic field be permanent, has long intrigued workers in this field. However, with some materials the superconducting current is carried only in a thin surface layer and the superconductivity is easily destroyed by the magnetic field. These materials, which include aluminium, tin, mercury and lead, are referred to as "soft" superconductors. On the other hand, "hard" superconductors, e.g. intermetallic compounds such as niobium-tin, niobium–titanium or vanadium–gallium, carry the current in thin filaments throughout the structure and are relatively resistant to magnetic fields.

With a hard superconductor the only power required, once the magnet is at full strength, is that required by the refrigerator used to keep the magnet cold, i.e. to compensate for heat influx. Apart from economy in power there is a vast potential reduction in capital cost because of the very small size of the magnet. An example which has been calculated is that of a magnet which produces 100,000 times the strength of the earth's magnetic field. A conventional electromagnet would have an iron core weighing many tons, together with a high current copper coil. Fifty kW of power would be required and most of this would be converted into heat which would have to be dissipated. An equivalent superconducting magnet would weigh only a few hundred pounds, including its refrigerator, and would require little more than the power used in refrigeration. Many hundreds of experimental and pilot scale superconducting magnets have been built and commercial application in the electrical equipment field may become commonplace.

A further important and promising application of superconductivity in the field of electrical equipment is for electric motors. Such motors would be essentially free of the internal electrical and mechanical losses of ordinary electric motors.

Still another application of superconductivity which has been considered is in the field of electrical transmission. Losses of the order of 15 per cent are incurred in present practice between the point of production of electricity and its point of use. If transmission lines could be kept at liquid-helium temperatures not only would this loss be greatly reduced but capital cost would be reduced because much lighter lines would be adequate. This is especially important if still greater amounts of power than at present can be transmitted. It seems possible that these savings would be greater than the cost of providing refrigeration at such a low temperature.

It is clear that if any of these ideas is to reach practical application an essential feature of the equipment will be helium refrigerators of absolute security, together with the ability to carry out maintenance work at liquid-helium temperatures.

OTHER LIQUID-HELIUM APPLICATIONS

Not all possible applications of liquid helium depend on superconductivity. There are some electronic devices which operate more efficiently at these low temperatures. An example of this is the "maser", a device for microwave amplification by stimulated emission of radiation. The material used for reception of signals in this way is a synthetic ruby, i.e. alumina containing controlled amounts of chromium ions. If the device is operated at liquid-helium temperature, interference from the thermal vibrations of the atoms in the crystal is eliminated.

An alternative device for the same purpose is the parametric amplifier. This contains a capacitance whose stored charge is not linearly related to the applied voltage, and when several alternating voltages are applied it can transfer energy from one frequency to another. In this way it can be used to amplify a signal received at low frequency. Again interfence is reduced by operating at temperatures down to that of liquid helium. Communication satellites depend on cryogenically cooled devices for the reception of signals, as for example the British station at Goonhilly in Cornwall.

Another device whose performance is improved at low temperatures is the infra-red detector of which the bolometer mentioned previously is an example. This is a much smaller and simpler device than radar and can perform much the same function. Further, it has no minimum operating distance and is difficult to jam. The lower the temperature at which it is kept the greater the sensitivity and range.

Finally it should be mentioned that liquid helium has a number of uses in nuclear and high-energy physics, especially for studying atomic and nuclear events without interference from the thermal motion of the atoms.

CHAPTER 10

Other Inert Gases

OF THE five inert gases first isolated by Ramsey we have seen that argon and helium are in common use and readily available in quantity. The other three, although much more difficult to isolate and therefore more expensive, have certain industrially valuable properties.

NEON

The least rare of the three is neon which is present in air in a concentration of 18 ppm. It will be seen from Table 1 in Chapter 2 that its boiling point is $-245 \cdot 9°C$, so that together with the helium in the air being processed it is not condensed at the temperatures used in air-separation plants. In the double column these two gases accumulate at the top of the condenser in the lower column, and can be withdrawn admixed with nitrogen gas and any traces of hydrogen present in the air (see Fig. 8.1). Even if this mixture is not withdrawn with the intention of recovering neon it must be vented from time to time to enable the air-separation plant to operate steadily and without drop in efficiency.

The mixture withdrawn contains about $2 \cdot 1$ per cent of neon and $0 \cdot 8$ per cent of helium. It is passed to a small column (Fig. 8.1) in which the bulk of the nitrogen is condensed out. The nitrogen content is then still further reduced by passing the product through a refrigerated carbon trap or through a liquid nitrogen bath at a pressure of 50–100 atm, leaving a product containing about 95 per cent of helium/neon. A further carbon trap removes the last traces of nitrogen, and passage over hot copper converts any hydrogen to water vapour which is then condensed out. Finally neon is separated from helium by adsorption on active charcoal at liquid-nitrogen temperature. The normal commercial grade of neon has a minimum purity of 98·0 per cent. The helium fraction is occasionally further purified for sale.

The chief use of neon is in electric signs. These are electric discharge lamps filled with particular gases or gas mixtures at 6–10 mm of mercury pressure. The electric current flows through the ionized gas and emits radiation characteristic of the gas in the lamp. Neon has its characteristic radiation mainly in the orange and red regions; these are regions which have the highest transmission in the worst weather conditions. In passing it may be mentioned that in an argon-filled lamp the radiation is blue, and that with various mixtures of neon, argon and helium colours of white, lavender, golden yellow and various shades of blue and green can be obtained.

Other uses for neon, usually in admixture with other gases, are in voltage-regulator tubes, Geiger counters, glow lamps, sodium vapour lamps and fluorescent tube starter switches.

KRYPTON

Krypton and xenon have boiling points higher than oxygen and tend to accumulate eventually at the base of the upper column in an air separation plant. In plants designed to make liquid oxygen the liquid is removed continuously and since it contains the krypton and xenon, accumulation of these elements is not important. On the other hand, in plants designed to make gaseous oxygen an equilibrium is reached in which the liquid oxygen contains as much as 50 ppm of krypton plus xenon against the 1·2 ppm present in air.

In order to recover the krypton and xenon this liquid is withdrawn (Fig. 8.1) and sent to a side column in which much of the accompanying oxygen is evaporated, leaving a product containing about 1 per cent of the inert gases. Any traces of acetylene or other hydrocarbons in the original air will by now have accumulated in the residual liquid. They are removed by evaporating the liquid and passing it over a catalyst at elevated temperature. The water vapour and carbon dioxide formed are then removed by a solid caustic-soda trap or by an absorbent. The two inert gases are now adsorbed on silica gel at low temperature; the product desorbed from the gel contains 90 per cent or more of inert gas. Finally by liquefaction and distillation or by adsorption and desorption on active charcoal pure krypton and pure xenon can be obtained.

The recovery of pure gases from such low initial concentrations necessarily involves these numerous purification stages and the cost of the product is relatively high. These gases are, however, valued for certain specific properties they possess and have been isolated industrially for many years.

The main uses of krypton lie in the electronic and lamp industries. In Chapter 8 it was mentioned that argon was preferred to nitrogen in gas-filled lamps partly because, on account of its higher molecular weight, it reduced both the thermal conductivity and the evaporation of the tungsten filament. Krypton has a much higher molecular weight and is still more effective. Indeed because of its poorer heat transfer properties, the size of the bulb can be reduced by as much as 50 per cent. The combined result is that a very bright light can be obtained from a relatively small bulb. Naturally the cost of such lamps is higher than the standard argon-filled ones but for special purposes, e.g. lamps on miners' helmets, this can be tolerated.

XENON

The method of isolating this gas has been described in the previous section. Because of its extreme rarity it has not found any wide use. However, its high density and inertness have led to it being used in scintillation counters. Its electrical properties are close to those of mercury and since it does not condense as readily it has been used to supplement mercury in some thyrotron tubes. It is even more effective than krypton in lamps. For example, if it is used as the filling gas a 300,000-candle-power lamp can be made which is only the size of a normal 40-watt bulb.

Xenon is a mild and non-explosive anaesthetic, but because of its rarity it is not used in practice.

CHAPTER 11

Acetylene

ACETYLENE is the first member of the acetylene series of hydrocarbons, or alkynes, which have the general formula $R_1C\equiv CR_2$. Because of the triple bond between the two carbon atoms these compounds are both endothermic and highly reactive. Thus the heats of formation of ethane, ethylene and acetylene at 15°C are:

Ethane	C_2H_6	$+ 20$ kcal/mole
Ethylene	C_2H_4	$- 12\cdot5$ kcal/mole
Acetylene	C_2H_2	$- 54\cdot2$ kcal/mole

One consequence of the endothermicity of acetylene is that its heat of combustion is very high, so that a flame with a very high temperature can be produced, especially if oxygen is used instead of air.

Both because of its high flame temperature and its great chemical activity acetylene is an important industrial gas and considerable effort has been devoted to developing methods of producing and handling it.

Since acetylene is so highly endothermic the equilibrium

$$2C+H_2 \underset{\longleftarrow}{\longrightarrow} C_2H_2$$

is almost entirely on the side of dissociation at temperatures in the range 0–2500°C. Consequently it must be made by an indirect method and in practice two methods have been employed. The first is to decompose calcium carbide with water and the other is to crack suitable hydrocarbons under controlled conditions. Each of these will be discussed in this chapter.

CALCIUM CARBIDE

In 1836 Edmund Davy announced that while trying to prepare potassium by calcining tartar with charcoal he had obtained a black solid which yielded a gas on reaction with water. He went on to describe its properties and called it "bicarburet of hydrogen". He even suggested it might be a useful illuminant. What he had produced was potassium carbide which reacts with water to give acetylene.

Berthelot in 1860 made acetylene by various more direct methods and gave it its modern name. Wohler in 1862 prepared calcium carbide by the reaction of carbon with a zinc–calcium alloy, and noted that it gave acetylene on reaction with water.

The large-scale production of calcium carbide, however, had to await the development of the electric furnace by Siemens in 1877. In May 1892 Willson in the United States was trying to make metallic calcium in order to use it to reduce alumina to aluminium. He was doing this by reacting coal tar and lime in an electric furnace. What he actually produced was calcium carbide by the reaction:

$$CaO + 3C \rightarrow CaC_2 + CO$$

In December of the same year Moissan in France, independently and on a smaller scale, also made calcium carbide. Once it was seen how this material could be made on a commercial scale development was rapid and by 1896 calcium carbide was an industrial chemical in several countries. Although the basic principles have remained the same over the years, the method of production has undergone considerable change, particularly with respect to the design of furnace and type of electrode.

A simplified design of a modern furnace is shown in Fig. 11.1. The electrodes are of the Soderberg type, i.e. a hollow electrode filled with pitch. As the electrode is consumed at the bottom the descending pitch is carbonized by the high temperature in the furnace, so forming a carbon electrode *in situ*. A mixture of coke and lime is fed to the top of the furnace and as it descends is warmed by heat exchange with the gases leaving the reaction zone. Three-phase current is employed at a potential of 75–250 volts and a current of 50,000–125,000 amperes. The resistance of the charge to the passage of the current causes it to heat up and reaction takes place at 2000–2500°C. Frequently it is arranged that

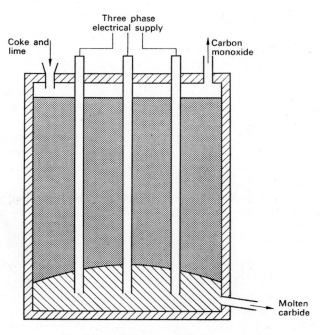

FIG. 11.1. Calcium-carbide furnace.

the furnace rotates slowly in order to prevent bridging of the charge.

The molten carbide is tapped into trucks where slow cooling to room temperature occurs. The cold material is broken into material of suitable size, graded, and stored in steel drums. Because it reacts so readily with moisture in the air it is important to avoid contact with the atmosphere as far as possible.

The theoretical power consumption is 2045 kWh per ton. Most commercial units require about 3000–3500 kWh per ton of 80 per cent carbide, equivalent to 3750–4375 kWh per ton of pure carbide. The reasons for the low efficiency are threefold: the products leave the furnace at high temperature: in order to achieve a reasonable melting temperature it must be arranged that the tapped carbide contains about 15 per cent of free lime; and impurities in the coke and lime react to give unwanted materials such as ferrosilicon, aluminium and magnesium. The net result is that the carbide is normally a little over 80 per cent pure, a figure which it is required to contain by British Standard Specification.

The annual world production of calcium carbide in 19 72 was about 4 million tons of which about 100,000 was produced in the United Kingdom. Not all of the world production was used to generate acetylene, since as described in Chapter 7 some is reacted with nitrogen to make calcium cyanamide.

ACETYLENE FROM CARBIDE

The generation of acetylene from calcium carbide by reaction with water takes place at room temperature:

$$CaC_2 + H_2O \rightarrow C_2H_2 + CaO$$

Although the reaction is so simple Lewes in 1900 described as many as thirty-seven different generators and listed 527 British patents for them. The reason for this great activity was that it was hoped that acetylene flames would be used to illuminate houses, factories and public buildings and that each would have its own generator. Although an acetylene flame is much brighter than an ordinary coal-gas flame, acetylene was in fact developed a little too late to compete with town's gas. The reason for this was that the Welsbach mantle, invented in 1886, enabled town's gas to be burnt under conditions in which a very bright light was obtained.

Methods of generating acetylene from carbide depend on one of two principles. In "wet" generation carbide is added to water as required and in "dry" generation water is added to carbide. At first sight it might be thought that the second principle would be preferred as exactly the amount of acetylene needed would be made by controlling the addition of water. While this is true in the case of very small-scale generation, say in a bicycle lamp, the issue is not so clear on an industrial scale. The exothermic heat of reaction is dissipated in the water in wet generators causing only a small temperature rise, while in dry generators a much larger rise occurs. One consequence is that different impurities are produced and those from dry generators are more difficult to remove. On the other hand, the sludge from wet generators is much more difficult to dispose of than the dry lime from dry generators. In practice the decision depends on a number of local factors, including the purpose for which the acetylene will be used.

The chief impurities in acetylene from wet generators are hydrogen sulphide, ammonia, phosphine and arsine. Gas from dry generators also includes small amounts of organic sulphur compounds. The impurities are removed by passing the gas through reagents such as chromic acid, acidified copper hypochlorite or ferric salts, which oxidize them and then absorb the products formed. The resulting gas is about 99·5 per cent pure, the main impurity being nitrogen.

STORAGE AND DISTRIBUTION

Acetylene is liable to explode at room temperatures and pressures above 1·4 atm if subjected to a modest initiating force.† Early attempts to store and transport acetylene as a liquefied gas under pressure led to a number of serious accidents. However, Claude and Hess in 1896 showed that acetylene was very soluble in acetone, e.g. 24 volumes per volume at 20°C and atmospheric pressure. Later in the same year Claude also showed that if the solution of acetylene in acetone is absorbed in a porous mass which completely fills the containing vessel, additional safety is obtained. This principle is still used generally for transporting acetylene gas in cylinders. Most industrial countries have strict statutory regulations about dissolved acetylene cylinders, including the maximum pressure, at a defined temperature, to which they may be filled. In the United Kingdom the figure is 225 psi gauge at 60°F (approximately 15 atm at 15°C).

Since the original observations by Claude and Hess many solvents for acetylene other than acetone have been examined. Dimethylformamide and *N*-methylpyrolidone absorb 34 and 29 volumes respectively at 20°C and atmospheric pressure. However, it must be remembered that the useful acetylene content of a cylinder is the difference between what was originally dissolved and what remains at the final pressure. It is

† Under different conditions various workers have obtained other results. See reference 13 in the Bibliography for a full account of this subject.

therefore necessary to study solubilities at pressures up to 15 atm in order to carry out the appropriate calculations. Further some solvent is lost by evaporation into the outgoing acetylene, and both the initial and the replacement cost of the solvent are important economic considerations. For these reasons acetone is still the preferred solvent in general use in cylinders.

The type of porous mass used in the cylinders varies. Some operators prefer a granular mass, such as kapok or a mixture of charcoal and kieselguhr, which can be poured into the cylinder and shaken down. Others prefer a monolithic mass, made for example by reacting sand and lime *in situ* followed by drying. Whichever method is used it is important that there should be a large number of minute pores into which the acetone can penetrate, and an even density throughout the cylinder.

Safety devices are frequently provided on each cylinder in case, despite the use of solvent and porous mass, decomposition of acetylene should start. In the United Kingdom a bursting disc is provided at each end of the cylinder so as to ensure that the pressure exerted by the products of decomposition cannot reach a figure high enough to burst the cylinder. An alternative system, used in some countries, is to incorporate in the cylinder wall plugs of low-melting-point alloys, which will melt if exothermic decomposition takes place and so release the gas.

Because of the system which has to be adopted a cylinder designed to carry 15 lb (200 ft^3) of gas weighs 195 lb without acetylene and 210 lb when full; the steel cylinder itself weighs 145 lb. Nevertheless in the United Kingdom about 16,000 tons of acetylene is distributed annually in cylinders. Users of acetylene who have a large demand usually install batteries of cylinders linked by manifolds to supply lines to the points of usage. Supply by pipeline from generators to the consumption site is almost entirely confined to chemical works.

Acetylene should never be allowed to come into contact with copper or any alloy containing more than 70 per cent of copper, since there would then be danger of the highly explosive copper acetylide being formed. All fittings and lines must be made of appropriate materials.

Since the cost of transporting acetylene in cylinders is higher than the cost of transporting the equivalent amount of carbide in drums, and the cost of manufacture of acetylene from carbide is nearly independent of scale, the most economic way of providing acetylene for general distribution is to transfer carbide from a central production plant to local acetylene generating and compression plants to supply a locality. Because local demand is likely to be variable, or even intermittent, the fact that acetylene generation from carbide is capable of intermittent operation is a considerable advantage.

ACETYLENE FROM HYDROCARBONS

A number of processes have been developed in recent years for making acetylene by the cracking of hydrocarbons. These are high-temperature processes which must be operated continuously and are most likely to be economic on a large scale, e.g. when acetylene is being made in bulk for steady use on a single site. The principle on which all these processes are based is to heat an aliphatic hydrocarbon feedstock rapidly to a temperature of 1000–2000°C or more, and rapidly cool the products formed. The whole process takes between 0·001 and 0·01 second. Since acetylene is unstable at high temperatures extremely rapid cooling is essential. Since the rate of decomposition of acetylene is proportional to its partial pressure it is advantageous to reduce the partial pressure either by working at reduced pressure or by dilution with steam or a combination of both.

The process of cracking is of the type

$$C_6H_{14} \rightarrow 3C_2H_2 + 4H_2$$

although other unsaturated compounds, e.g. ethylene, are also likely to be formed. It will be seen that the principal product must necessarily be hydrogen, and in any practical process the acetylene content in the product lies between 5 and 20 per cent. Many processes have been devised but few have been commercially successful. They differ from one another primarily in the means adopted to achieve the necessary high temperature, and can be divided into three types.

Firstly, there are electric arc processes which, although they have been technically successful, have been found to have an electric power consumption (13,000 kWh) per ton of acetylene which is greater than that required (10,000 kWh) to produce the same amount of acetylene by the carbide route.

The second method of rapidly heating a gas is in a flame. If a hydrocarbon is burnt in a deficiency of oxygen some acetylene is formed. Suitable feedstocks for this process are methane and ethane. Naturally a great deal of the gas is converted to carbon monoxide and hydrogen and indeed the economics are only favourable if this mixture can be used for chemical synthesis, for example of methanol. The outgoing gases can be used to preheat the hydrocarbon feedstock and the oxygen, in order to get a higher flame temperature and greater efficiency. Even then the product is still primarily carbon monoxide and hydrogen together with 8–9 per cent of acetylene, a few per cent of carbon dioxide and some unreacted feedstock.

The third process depends on the use of refractory heat exchangers in a cyclic process. The refractory is first heated to the temperature at which the feedstock will crack, by burning fuel in the furnace. The feedstock is then passed through and cracking takes place. The refractory cools down because of the transfer of heat to the feedstock and also because of the endothermic nature of the cracking reactions. The cycle time is of the order of 1 minute. Any soot formed during cracking is burnt off during the next combustion period. In order to minimize decomposition of the acetylene formed its partial pressure is kept low either by steam dilution or by operation at 0·5 atm pressure. In this process better yields are obtained with higher paraffinic hydrocarbons than with methane. In the United Kingdom three Wulff plants operating on this principle have been built. The process can be adapted so that it makes some ethylene as well as acetylene. In practice the manufacturer can vary the acetylene/ethylene ratio over the range 4/1 to 1/4.

It will be realized that the isolation of acetylene from the cracked products presents a major problem. The gas contains not only hydrogen, carbon monoxide and nitrogen but also higher acetylenes, olefines and paraffins. The most economic method is to employ a selective solvent such as acetone or dimethylformamide. In order to conserve the solvent, operation under high pressure or low temperature can be adopted. The extent of the purification and the precise method adopted depend on the use to which the acetylene will be put. Broadly it can be assumed that the cost of purification will be at least as great as the cost of the cracking stage. If purified to 99·5 per cent the impurities tend to be higher alkynes and not, as in the case of carbide acetylene, nitrogen. This fact alone would make the use of the gas for compression into cylinders unattractive, since these higher alkynes would tend to accumulate in the acetone solution.

Acetylene made by a hydrocarbon route is normally used for chemical production by a continuous process. Out of a total world output of about 2·4 million tons of acetylene in 1972 about 1·1 million tons were made from hydrocarbons.

USES OF ACETYLENE

As mentioned above acetylene was first produced as an illuminating gas but could not compete with town's gas. It still has a very limited use in lighthouses, lightships and buoys.

The oxy-hydrogen flame was well known before acetylene became readily available, but by 1895 Le Chatelier was experimenting with the hotter oxy-acetylene flame. The experimental determination of flame temperatures requires special techniques. They can also be calculated theoretically providing it is borne in mind that both carbon dioxide and water vapour are partially decomposed at equilibrium at elevated temperatures:

$$2CO_2 \rightleftarrows 2CO + O_2$$
$$2H_2O \rightleftarrows 2H_2 + O_2$$

Each of these equilibria moves to the right, endothermically, as the temperature is raised. Evidently in any practical case less oxygen has to be supplied than would be the case if there were no dissociation. In fact it is found that the highest theoretical temperature is obtained when the ratio of oxygen to acetylene is approximately 1/1 and not 2·5/1 as simple stoichiometry would suggest. The theoretical flame temperature at normal atmospheric pressure is 3114°C. The average flame temperature actually measured is somewhat below this owing to heat transfer losses, although that measured in the centre of the flame by the sodium line reversal method is only slightly below this value. For the comparison it may be noted that if air is used instead of oxygen the maximum theoretical flame temperature is 2300°C.

The two main operations involved in metal treatment and shaping are cutting and joining. The joining of metals can be undertaken in various ways such as bolting or riveting, but until Fouche and Picard developed the oxy-acetylene blowpipe in 1901 the only method of welding was "forge" welding. The development of this blowpipe enabled fusion welding to be carried out with greater ease and speed than forge welding. A highly localized flame is produced which minimizes distortion of the metal and produces a free flow of metal to form the weld. Gas welding is not suitable for repetitive processes and a variety of electric welding processes have been developed which find wide application in industry. Nevertheless, oxy-acetylene welding still finds application where intermittent use is made of welding. The equipment is cheap and easily moved; welding can be done in all positions; it can be used for both ferrous and non-ferrous metals; and the finish is fair although the speed is lower than for some other processes. It is suitable for metal thicknesses up to about half an inch, and is used in general jobbing and repair work.

For welding steel the acetylene and oxygen flows from the cylinders are adjusted to be equal in volume and since complete oxidation does occur, as mentioned earlier, the reducing gases protect the weld metal from oxidation by the surrounding air. When welding brass an oxidizing flame is used because this produces a layer of zinc oxide on the surface and this reduces the loss of zinc by volatilization. On the other hand, when applying certain hard facing coatings to a metal or when welding metals which oxidize very readily, an acetylene-rich flame is an advantage.

The use of acetylene in the oxygen-cutting process has been discussed in Chapter 6.

CHEMICAL USES OF ACETYLENE

Acetylene is extremely reactive and potentially could be the starting point for many chemical products. However, only a small number of such chemicals have any commercial value and in practice the only ones to have been manufactured from acetylene on any scale have been acetaldehyde, vinyl chloride, vinyl acetate, acrylonitrile, acrylic esters, trichloroethylene, neoprene and acetylene black. Of these the first six can also be made, perhaps not always so directly, from ethylene, which is available at a price only 40 per cent of that of acetylene. There is a tendency, therefore, for processes which were based on acetylene when ethylene was relatively dear, to be replaced by ones based on ethylene or other olefines.

A brief account is given below of the method of production of each of the eight materials mentioned above.

Acetaldehyde

If acetylene is passed countercurrent to a sulphuric acid solution containing mercuric sulphate, the acetylene is hydrated to acetaldehyde:

$$C_2H_2 + H_2O \rightarrow CH_3 \cdot CHO$$

Acetaldehyde is a chemical intermediate, the majority of it being converted to acetic acid and acetic anhydride by oxidation.

Vinyl Chloride

This material is made by the direct reaction of acetylene with hydrogen chloride at 100–180°C over a catalyst such as mercury chloride supported on charcoal:

$$C_2H_2 + HCl \rightarrow CH_2 : CHCl$$

On polymerization vinyl chloride gives polyvinylchloride (PVC) which is widely used for articles such as electric cables, lightweight macintoshes and floor tiles.

Vinyl Acetate

This material is made by the direct reaction of acetylene with acetic acid in the vapour phase at 170–210°C over a catalyst of zinc acetate on charcoal:

$$C_2H_2 + CH_3 \cdot COOH \rightarrow CH_2 \cdot COO \cdot CH:CH_2$$

On polymerization polyvinylacetate is formed; its principal outlet is in emulsion paints in which pigments are dispersed in an aqueous emulsion of the polymer. It is also used in copolymers and for making polyvinyl alcohol and polyvinylacetals.

Acrylonitrile

Another name for this material would be vinyl cyanide and it is made by the direct reaction of acetylene with hydrogen cyanide:

$$C_2H_2 + HCN \rightarrow CH_2CHCN$$

A large excess of acetylene is used and the gases are circulated at 80–90°C through a hydrochloric acid solution of cuprous chloride which acts as the catalyst. The first use of acrylonitrile was as a co-monomer with butadiene in the manufacture of oil-resistant rubbers. It is now used for polymerization either alone or with smaller amounts of other monomers to make a variety of fibres, e.g. Orlon, Chemstrand and Acrilan.

Acrylic Esters

These can be made by the reaction of acetylene with carbon monoxide in the presence of the appropriate alcohol, together with nickel carbonyl and hydrochloric acid. The reaction is carried out at atmospheric pressure and a temperature of 30–50°C:

$$C_2H_2 + CO + ROH \rightarrow CH_2:CH \cdot COOR$$

These materials are mainly used in homo- or co-polymers.

Trichloroethylene

This material is made in two stages. Firstly acetylene is reacted with chlorine at about 80°C using antimony chloride or ferric chloride as catalyst:

$$C_2H_2 + 2Cl_2 \rightarrow \begin{array}{c} CHCl_2 \\ | \\ CHCl_2 \end{array}$$

This tetrachlorethane is then treated with a lime slurry at its boiling point, or at 230–320°C over a barium catalyst, in order to remove hydrogen chloride :

$$\begin{array}{ccc} CHCl_2 & & CCl_2 \\ | & \rightarrow & \| & +HCl \\ CHCl_2 & & CHCl \end{array}$$

Trichloroethylene is an excellent degreasing solvent widely used in dry cleaning and other operations. Since it has a reasonably low vapour pressure at room temperature losses by evaporation are small.

Neoprene

This material is a synthetic rubber which has been manufactured in the United States for over 20 years and production exceeds 100,000 tons per annum. United Kingdom production started in 1960 in Northern Ireland. Three stages are involved. The acetylene is first converted to vinyl acetylene:

$$2C_2H_2 \rightarrow CH_2 : CH \cdot C : CH$$

Reaction with hydrochloric acid then gives 2-chlorobutadiene or chloroprene:

$$CH_2 : CH \cdot C : CH + HCl \rightarrow CH_2 : CH \cdot CCl : CH_2$$

The chloroprene is then polymerized to give neoprene.

Acetylene Black

Acetylene black is obtained by the controlled decomposition of acetylene into its elements in a retort at about 800°C. The reaction is, of course, highly exothermic, and once started is thermally self-supporting. The product is a very pure form of carbon (over 99·5 per cent). It has special electrical properties not shown by other forms of carbon black and is mainly used in the manufacture of dry batteries. It is also used in rubber and plastics compounding to produce an antistatic, or conductive, material. The principal production is in Canada where about 10,000 tons per annum are made.

CHAPTER 12

Hydrogen

HYDROGEN was the first gaseous element to be isolated in a reasonably pure state. This was achieved by Cavendish in 1766 by the reaction of dilute sulphuric acid and zinc. Today it has a number of very important industrial uses, either alone or mixed with other gases. The method of production employed depends on the quantity and purity required and on considerations regarding the availability and cost of raw materials.

PRODUCTION

The simplest way of making hydrogen is by the electrolysis of an appropriate aqueous solution. This method is normally used only when the scale of operation is small (up to 100–500 tons per annum) or the demand is intermittent, since under other conditions the cost of power in most countries makes the method uneconomic. It should be noted that the electrolysis of brine to produce chlorine and caustic soda also results in the production of the same volume of hydrogen as chlorine:

$$NaCl \rightarrow Na^+ + Cl^-$$
$$2Cl \rightarrow Cl_2$$
$$2Na + 2H_2O \rightarrow 2NaOH + H_2$$

It is not always possible to use the by-product hydrogen except as a fuel.

Another method for making hydrogen from water is to react steam with spongy iron at about 650°C and atmospheric pressure:

$$3Fe + 4H_2O \rightarrow Fe_3O_4 + 4H_2$$

The iron oxide formed can be reduced back to the metal by blue water gas. This method is no longer used in the United Kingdom.

Reference has already been made (Chapter 6, p. 45) to the production of blue water gas and the further reaction of the carbon monoxide with steam to give carbon dioxide and hydrogen. If the carbon dioxide is removed by a suitable absorbent, hydrogen, not highly pure, is left.

Natural gas can be converted to carbon monoxide and hydrogen by reaction with steam, for example at 900°C over a nickel catalyst promoted by magnesia or alumina:

$$CH_4 + H_2O \rightarrow CO + 3H_2$$

The carbon monoxide can be reacted with more steam, as in the water gas shift reaction, so that the overall reaction becomes

$$CH_4 + 2H_2O \rightarrow CO_2 + 4H_2$$

It will be seen that half of the hydrogen comes from the natural gas and half from the steam. These reactions are used industrially to manufacture hydrogen and synthesis gas for methanol production or the Fisher–Tropsch reaction. Higher hydrocarbons can be used in place of methane and the equilibrium is then slightly more favourable. Clearly such raw materials give a higher ratio of carbon monoxide to hydrogen in the product.

The reaction between the hydrocarbon and steam is endothermic and the temperature in the converter must be maintained either by an external source of heat or by reacting part of the hydrocarbon with oxygen internally. In the latter case the thermal balance is maintained by using a steam/oxygen mixture of such a ratio that the endothermic reaction with steam plus the heat losses from the system are just balanced by the exothermic reaction with oxygen.

CONSUMPTION

The total United Kingdom consumption of hydrogen in 1968 was about 500,000 tons, of which over half went into synthetic ammonia manufacture, and much of the rest into methanol and nylon production and for refinery uses. About 16,000 tons was sold by the industrial gas companies in cylinders. Depending on customer requirements the purity was either 99·9 per cent or 99·99 per cent. The main uses are given in Table 6.

TABLE 6. INDUSTRIAL HYDROGEN USES

Use	Tons per annum	per cent
Chemical industry	9280	58
Food	3840	24
Metallurgical	1920	12
Lamps and glass	480	3
Electronics	320	2
Miscellaneous	160	1
Total	16,000	100

USES

Apart from the uses of hydrogen in major chemical processes such as reaction with nitrogen to give ammonia, and with carbon monoxide to give methanol, it is used to convert olefines to aldehydes by the OXO process:

$$C_2H_4 + CO + H_2 \rightarrow CH_3 \cdot CH_2CHO$$

This reaction is carried out in the presence of a cobalt catalyst at temperatures of 110–180°C and pressures of 100–200 atm.

As will be seen from Table 6 hydrogen has a wide variety of uses in industry. It is used on a large scale in the petroleum industry, e.g. for the conversion of naphtha to motor spirit by hydroforming. The hydrogenation of unsaturated materials to saturated ones in the presence of a nickel catalyst is known as "fat hardening" and is a key feature in the manufacture of margarine.

In the production of metals there are several instances of the direct reduction of the oxide to the metal by the use of hydrogen, e.g. germanium and tungsten. In metal fabrication the oxy-hydrogen flame is used in place of the oxy-acetylene flame in welding certain low-melting-point metals. The so-called atomic hydrogen welding process depends on an electric arc struck between two tungsten electrodes and maintained in an atmosphere of hydrogen. The hydrogen molecules dissociate into atoms which recombine on the surface of the work with the liberation of a large quantity of heat. This process has been used for certain special applications such as welding of high chromium–nickel steels.

Hydrogen is used for cooling alternators in the electricity industry because of the good heat transfer resulting from its high rate of diffusion. It is sometimes used for filling balloons despite its flammability.

SYNTHETIC AMMONIA

Virtually all ammonia produced today is made synthetically by the reaction of hydrogen and nitrogen over a catalyst at high pressure and temperature:

$$3H_2 + N_2 \rightleftharpoons 2NH_3$$

The higher the pressure the more the equilibrium moves to the right and the higher the temperature the more it moves to the left. It is important, therefore, to operate at as low a temperature as possible, say around 500°C. At this temperature the rate of reaction is very slow and the use of very active catalysts is essential for an adequate conversion rate to be achieved. Many of the original researches were carried out in the years before World War I by the German scientist Haber. Based on his work Bosch designed the first commercial plant, which commenced operation in 1913. Since that time the Haber–Bosch process has been improved and a number of alternative processes developed. Except for the Claude process, which operates at 900–1000 atm, conversions are such that gas recirculation is essential.

The current Haber–Bosch process operates at 200–300 atm and 500°C to give conversions of 20–22 per cent per pass. The catalyst, as in most of the other processes, is promoted iron. Indeed it is doubly promoted, firstly by an acidic or amphoteric oxide such as alumina, and secondly by an alkali oxide such as that of potassium. Iron oxide is mixed with the promoters and fused electrically. It is then reduced to iron by the action of hydrogen.

The hydrogen, made by one of the methods described above, can be mixed with nitrogen made by air liquefaction; or alternatively some of the hydrogen can be burnt in air so as to give a residual gas of the right composition after removal of water vapour. Another possibility is to add producer gas to blue water gas before reacting the carbon monoxide with steam to bring about the water gas shift reaction (see p. 45). Whichever method is used it is essential for the gas mixture to be purified not only from all sulphur compounds but also from carbon monoxide to below 10 ppm and carbon dioxide to below 15 ppm.

Space velocities of the order of 20,000 to 40,000 volumes of free gas per volume of reactor space per hour are achieved on modern plants. Since the reaction is exothermic and overheating can cause loss of catalyst activity, special designs of converter are used to ensure adequate heat removal.† The ammonia leaving the plant is either condensed out or absorbed in water. In the former case pressures of at least 2·7 atm are used; the boiling point of ammonia at this pressure is −3·5°C so that refrigeration is necessary. In order to keep transport costs as low as possible bulk quantities of ammonia are shipped in the anhydrous state.

Up to 70 per cent of the ammonia made in the United State goes for agriculture in one form or another, e.g. as sulphate, phosphate or nitrate. Ammonia is also used on a large scale for conversion to nitric acid, for the production of soda ash from salt, and for explosives and urea. On a somewhat smaller scale it has a wide variety of uses, e.g. in refrigeration, for cracking to give an inert atmosphere (see p. 49), as a cleaning agent and in the manufacture of pharmaceuticals.

LIQUID HYDROGEN

The only large-scale use of liquid hydrogen is in rocketry where it is preferred to other fuels because its low density results in a high velocity in the exhaust gases emerging from the rocket nozzles. Hydrogen gives the highest specific impulse of all known fuels (see also p. 47). Production of liquid hydrogen on a large scale is therefore confined to the United States and presumably Russia. It can be effected by using the same principles as for the liquefaction of air, although there are three special difficulties which should be noted. Firstly, since the inversion temperature of hydrogen is −69°C, it must be cooled below this temperature

† For further details see reference 14 in the Bibliography.

before use can be made of the Joule–Thomson effect. In practice this is done by cooling with liquid nitrogen.

In the second place it should be noted that all substances, other than helium and hydrogen, freeze above −253°C which is the boiling point of liquid hydrogen. Since they would then cause a blockage in the plant it is essential to remove all impurities completely, e.g. by adsorption on silica gel or charcoal. Particularly serious is the presence of small amounts of oxygen, and explosions in hydrogen liquefiers have been attributed to the accumulation of solid oxygen in tubes carrying liquid hydrogen.

The third problem arises from the fact that hydrogen consists of two molecular varieties, distinguished by the relative orientations of the nuclear spins in the diatomic molecule as illustrated in Fig. 12.1. The

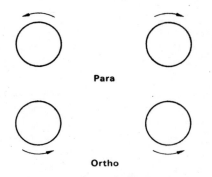

Para

Ortho

Fig. 12.1. Ortho and para-hydrogen.

ratio of the two forms present in any sample of hydrogen at equilibrium depends on the temperature. At room temperature the ortho/para ratio is 3/1, but at the boiling point of liquid hydrogen it is 1/500. The rate of conversion of ortho-hydrogen to para-hydrogen is rather slow but it is exothermic. Consequently if the hydrogen when liquefied has a concentration of ortho-hydrogen greater than the equilibrium amount the exothermic reaction will proceed and result in the evaporation of some of the liquid. It is essential therefore to complete the ortho–para conversion, using catalysts such as iron oxide, at liquid-hydrogen temperature before sending the liquid to storage.

Another possible use for liquid hydrogen may arise if thermonuclear reactions involving deuterium or tritium as fuels become a practical reality. Deuterium occurs in natural hydrogen to the extent of one part in 6900. It can be separated by distillation of liquid hydrogen. Alternative methods are the distillation of water and catalytic exchange between water and water vapour.

CHAPTER 13

Carbon Dioxide

CARBON dioxide is the final product of many reactions and is available in impure form in quantities greatly in excess of demand. The main technical problems therefore lie not in its production but in its separation and storage and distribution.

It occurs as a by-product on a large scale from four types of process:

1. In flue gases from the combustion of carbonaceous fuels in air, accompanied by large quantities of nitrogen.
2. In the manufacture of hydrogen by the reaction of steam with a carbonaceous materials as mentioned in Chapter 12.
3. From lime kilns in which calcium carbonate is decomposed to the oxide.
4. In the fermentation of molasses, corn, wheat, potatoes and other materials to make alcohol. The basic process is the conversion of a sugar such as dextrose by the reaction:

$$C_6H_{12}O_6 \rightarrow 2C_2H_5OH + 2CO_2$$

In addition carbon dioxide is a by-product of other processes such as the manufacture of sodium phosphate by the reaction between sodium carbonate and orthophosphoric acid. It is also found in considerable concentrations in some natural gas wells. It is, however, rarely recovered from such wells as they are usually at isolated sites remote from the point of demand.

RECOVERY

If the source of carbon dioxide is flue gases or hydrogen production processes it can be stripped out by dissolving in a solvent such as ethanol-amine at a temperature of 25–65°C. The solution is then removed and heated to 100–150°C to release the carbon dioxide for collection. Gas from lime kilns must first be treated to remove dust. This is usually done in two stages, firstly by passage through a dry chamber where most of the entrained dust settles, and then by washing with water. Gases from the above sources are liable to contain a little sulphur dioxide or hydrogen sulphide. The taste and odour of the latter would be quite unacceptable in carbon dioxide used for beverages or dry ice. It can be oxidized by potassium permanganate or dichromate followed by removal of the oxides and sulphur so formed. Alternatively impurities can be removed by adsorption on active charcoal.

The by-product gas from fermentation processes contains a higher concentration of carbon dioxide than the other sources. It is purified either chemically or by adsorption on active carbon.

STORAGE AND DISTRIBUTION

Carbon dioxide has one peculiar physical property which complicates its handling. If cooled at atmospheric pressure it passes straight from the gaseous to the solid state at −78·5°C. It can be obtained as a liquid only under pressure. All three states coexist at the triple point which is −56·6°C and a pressure of 5·11 atm. The critical temperature for carbon dioxide is 31°C and its critical pressure is 72·85 atm. It can, therefore, be obtained in the liquid form only between −56·6°C and 31°C under the appropriate pressure. Carbon dioxide is normally stored industrially in liquid form, the technique adopted being to compress the gas to about 75 atm and then cool with water to make sure that the temperature falls below the critical one. Sometimes it is stored at a lower temperature, e.g. −13 to −23°C, in which case it need only be compressed to 15–23 atm. The compressed gas is first cooled to room temperature when impurities such as water vapour or entrained oil can be removed. The gas then passes to a refrigerant cooled condenser where it liquefies.

Modest quantities (say up to 50 lb) are conveniently transported as liquid in cylinders at ambient temperature. Larger quantities are both transported and stored at the customer's site in refrigerated and insulated tanks maintained at about $-18°C$ and 20 atm. Storage tanks at the customer's works may hold 2 to 50 tons and are kept cool by a Freon refrigeration unit.

Until recently most customers were supplied with carbon dioxide in the solid form referred to as "dry ice". If carbon dioxide in a cylinder is allowed to expand to atmospheric pressure a yield of about 25 per cent of snow can be obtained. This method is only suitable when the customer requires small, intermittent amounts of solid. Larger customers receive the solid in blocks made by the supplier who feeds liquid into a press chamber at $-45°C$ and a pressure just over 5·1 atm. On reducing the pressure a mass of solid is produced which can be compressed into a solid block by hydraulic rams. Losses during the transport and distribution of dry ice are about 10 per cent.

USES

Dry ice is used to refrigerate foodstuffs at a considerably lower temperature than is possible with ordinary ice; it also has the advantage of warming to become vapour and not liquid. It is also used as a coolant for a number of special purposes such as chilling golf-ball centres before winding, or chilling aluminium rivets prior to use. It is widely used as a convenient laboratory refrigerant.

The largest single use of gaseous carbon dioxide is for making carbonated beverages, including soda water. While production of these materials is largely carried out in industry, small cylinders of carbon dioxide (Sparklet bulbs) are sold to enable the drinker to make his own syphon of soda water.

It is used in fire extinguishers where it has the advantages of excluding air, acting as a coolant and tending to remain at floor level. It is used to "foam" rubber and plastics, and sometimes as an ingredient in the shielding gas in the welding processes described in Chapter 8.

An example of the use of carbon dioxide as liquid is in the Cardox blasting system. A cartridge of carbon dioxide is heated until the container ruptures, releasing a large quantity of gas with explosive violence.

A use of carbon dioxide which depends on a chemical reaction is in the hardening of sand cores in moulds. Sand and sodium silicate are poured into position and the gas introduced. The sodium silicate is converted into a mixture of sodium carbonate and bicarbonate together with silicic acid and sets to a hard mass.

The total size of the United Kingdom market for carbon dioxide in 1968 was about 200,000 tons.

CHAPTER 14

Miscellaneous Gases

IN THIS chapter a brief account will be given of a number of gases produced industrially but often distributed on only a small scale. Some of them are made on a much larger scale for captive use and the reader should refer to an appropriate textbook for more detailed information.

NITROUS OXIDE

Nitrous oxide, N_2O, was one of the first gases to be isolated. Although this was done by Priestley in 1772 it was Davy in 1794 who first noted its anaesthetic properties. The first published practical application of its use for this purpose was in Hartford, Connecticut in 1844. Despite the development of more potent gaseous anaesthetics and of intravenous anaesthetics it is still used. It is non-toxic, non-explosive and has no unpleasant after effects, but is only suitable for operations lasting thirty to ninety seconds as in dentistry. It is administered with oxygen in ratios of the order 90:10 to 95:5 nitrous oxide to oxygen. It might be noted that all inhalant anaesthetics other than nitrous oxide are volatile liquids which are supplied in liquid form and evaporated in inhalers. These include ether, chloroform, cyclopropane and trichloroethylene (trilene).

Nitrous oxide can also be used as a propellant for whipped cream, tomato sauce etc. because it has a suitable boiling point, a sweet taste, no toxicity and a low solubility.

Nitrous oxide is made by the action of heat on ammonium nitrate at a temperature in the range 200–260°C :

$$NH_4NO_3 \rightarrow N_2O + 2H_2O$$

Even with careful temperature control traces of other oxides of nitrogen and of ammonia tend to be formed and must be removed, e.g. by ferrous sulphate, caustic potash or milk of lime.

The nitrous oxide, which has a purity of at least 99 per cent, is compressed into cylinders. Since its critical temperature is 36·5°C and its critical pressure is 71·66 atm it is partly present in the liquid state but the material released from the cylinder is gas. The pressure in the cylinder remains constant as long as any liquid is present and then begins to fall; at normal room temperature this pressure is about 43 atm. Cylinder sizes are in the range 16 to 640 ft³ of gas. About 40 million ft₃ is used in the United Kingdom per annum.

LIQUEFIED PETROLEUM GASES

This term covers those petroleum hydrocarbons which, although gaseous at room temperature and pressure, are easily liquefied under pressure, and can therefore be conveniently stored and transported as liquid in light pressure vessels. The compounds which fall into this category are the C_3 and C_4 paraffins and olefines whose properties are given in Table 7.

TABLE 7.   LIQUEFIED PETROLEUM GASES

Compound	Formula	Boiling point, °C	Vapour pressure at 21°C, atm
Propane	C_3H_8	—42	8
Propylene	C_3H_6	—44	10
n-Butane	C_4H_{10}	— 1	2
iso-Butane	C_4H_{10}	—12	3
Butylenes	C_4H_8	— 7 to 1	2 to 3

Most liquefied petroleum gas is obtained from cracking and reforming operations, from distillation of crude oil at a refinery, or from wet natural gas from which it is separated by adsorption. In most countries two grades are sold, one of which is primarily propane and one which is primarily butane. In the case of mixtures rich in butane natural evaporation may not be sufficient to provide gas at a high continuous rate and a vaporizer may be required. For this reason propane is often preferred. Seventy to eighty per cent of the market for liquefied petroleum gas lies in the fields of cooking, water heating and space heating both in the industrial and domestic markets.

During the description in Chapter 6 of the process for the oxygen cutting of steel it was mentioned that propane could be used as the fuel instead of acetylene. The maximum theoretical temperature in an oxy-propane flame is 2826°C at an oxygen : fuel ratio of about 4·5 : 1, against 3114°C at a ratio of about 1:1 in the case of acetylene. This difference in temperature is sufficient to prevent propane being used in the gas welding of steel. In the case of cutting, however, the purpose of the fuel gas is to raise the temperature to that at which ignition of steel in oxygen will occur (1250°C) and thereafter to compensate for heat losses. Propane is perfectly capable of performing these functions and is widely used in this process. It should be noted, however, that the time required to preheat the metal and oxygen to the temperature at which the cut can be started is longer. Further on account of the different fuel to oxygen ratio needed much more oxygen is used per unit of heat produced. This partly offsets the saving in using the cheaper fuel gas. Indeed it is not possible to draw a general economic conclusion about the relative costs of using the two fuels, and each case must be separately costed taking into account labour charges and overheads as well as the equipment utilization time. Sometimes the fact that gas welding operations are also conducted on the same site affects the conclusion.

It should also be mentioned that propane cylinders are light in weight compared with acetylene cylinders. In the latter case (see Chapter 11) a cylinder yielding 200 ft^3 of gas weighs 210 lb when full and 195 lb when empty. In the case of propane a cylinder yielding 206 ft^3 weighs 61 lb full and 37 lb empty.

Finally the rapid rate of growth of the United Kingdom market for liquefied petroleum gases is shown by the following approximate figures:

1938   2400 tons

1948	19,000 tons
1959	126,000 tons
1963	450,000 tons
1971	1,200,000 tons

METHANE

Methane, CH_4, occurs naturally or is produced in an impure form industrially in a number of ways. It was first observed by Priestley in 1772 that it was formed during the gradual decay of vegetable matter. It is now known that this process proceeds best in the absence of air. When the gas is formed in ponds containing rotting organic material it is referred to as "marsh gas". In the disposal of sewage by the action of anaerobic bacteria up to 1 ft³ of methane is produced per person per day, together with about half as much carbon dioxide. This gas is sometimes collected for use as fuel at the sewage works, or even for compression into cylinders. In the latter case it is common practice to dissolve out the carbon dioxide by washing with water under pressure.

In coal mines an explosive gas called "firedamp" is sometimes found in pockets. This gas is rich in methane, but with one or two exceptions it has not been found possible to collect a steady stream of the gas in sufficiently pure form for practical use.

Gases such as coal gas or coke oven gas manufactured from coal by carbonization contain 20–50 per cent of methane depending on the manufacturing method employed. Owing to its high calorific value it contributes 40–50 per cent of the calorific value of town's gas, and it is necessary for any gas made by the complete gasification of coal to contain sufficient methane, as indicated in Chapter 6.

Natural gas frequently discovered in drilling for oil consists primarily of methane. Sources of natural gas in enormous quantities continue to be found all over the world and natural gas pipe lines have been built in many countries, sometimes up to several thousands of miles in length. Until recently no natural gas in any appreciable quantity had been found in or near the United Kingdom. The Gas Council therefore arranged to ship methane from North Africa in liquid form. The technical problems of liquefaction and storage in North Africa, of the design of suitable ships, and of methods of transfer of the liquid and evaporation after arrival in the United Kingdom, were all solved satisfactorily. However, subsequent finds of natural gas off the east coast of the United Kingdom have made this method of obtaining methane less economic. Plans are now in hand for the complete replacement of town's gas by natural gas by 1975. Since the calorific value of the latter is about twice that of the former this is necessitating conversion of all gas-using appliances.

Apart from its use as a fuel methane can be used for making synthesis gas (Chapter 6), acetylene (Chapter 11) and also carbon black by various thermal reactions. Although methane is the least reactive of all the hydrocarbon gases it can nevertheless be made to react with many materials. Thus with increasing amounts of chlorine it is possible to make methyl chloride, methylene chloride, chloroform and carbon tetrachloride:

$$CH_4 \rightarrow CH_3Cl \rightarrow CH_2Cl_2 \rightarrow CHCl_3 \rightarrow CCl_4$$

Incomplete oxidation of methane can be controlled to produce methyl alcohol and formaldehyde:

$$CH_4 \rightarrow CH_3OH \rightarrow H \cdot CHO + H_2O$$

Methane reacts with ammonia to give hydrazine; with sulphur or hydrogen sulphide to give carbon disulphide; and with nitric acid to give nitromethane.

ETHYLENE

The growth of the heavy organic chemical industry since World War II has to a great extent depended on the development of methods of producing ethylene at low cost and converting it into a range of useful materials such as polyethylene, ethylene oxide and glycol, ethanol and various chlorinated products. Indeed with ethylene capable of being produced at about 40 per cent of the cost of acetylene a number of products previously based on acetylene are now tending to be made from ethylene (see Chapter 11). The United Kingdom consumption of ethylene is approaching 1·4 million tons per annum in 1972.

Ethylene finds a use in fruit ripening and for this purpose is distributed as a gas under pressure in cylinders. For local reasons, such as the scale of operation and the predominance of distribution costs, it may be convenient, and indeed economic, to produce the ethylene intermittently as required for cylinder filling by the dehydration of alcohol:

$$C_2H_5OH \rightarrow C_2H_4 + H_2O$$

The alcohol is passed at a temperature of about 350°C over a catalyst which may be alumina or phosphoric acid on coke. After a time a certain amount of carbon deposits on the coke and has to be removed. The ethylene is liable to contain traces of acids and aldehydes which can be removed by scrubbing the gas with water.

The controlled ripening of various fruits and vegetables, particularly citrus fruits, is of considerable commercial importance. Ethylene is a ripening hormone and widely used.† Ethylene also has anaesthetic properties, but these are only of the same order as those of nitrous oxide, and it is no longer used for this purpose.

CARBON MONOXIDE

It has been mentioned in Chapter 12 that vast quantities of carbon monoxide are used in association with hydrogen for certain chemical synthesis. It is also required in large amounts for the manufacture of nickel. After reduction of the oxide to the metal the nickel is purified by reacting it with carbon monoxide at about 45°C to give nickel carbonyl:

$$Ni + 4CO \rightarrow Ni(CO)_4$$

The gaseous carbonyl is then decomposed to pure nickel and carbon monoxide for recycling by passing it over nickel pellets at about 180°C. Originally these reactions were carried out at atmospheric pressure but in recent years pressures of 20 atm have been used.

A number of esters, acids and hydroxyacids, e.g. acetic acid, propionic acid, methyl acetate, methyl formate and ethyl propionate, are made by the reaction of 99 per cent pure carbon monoxide with appropriate

† For a fuller account see reference 16 in the Bibliography.

materials. One example is the reaction with ethylene and water at a pressure of 200 atm and a temperature of 250–285°C in the presence of a nickel salt catalyst. Propionic acid is formed by the reaction:

$$C_2H_4 + CO + H_2O \rightarrow C_2H_5 \cdot COOH$$

The carbon monoxide for this reaction is made by the selective absorption of the material from water gas or some similar gas, using, for example, cuprous ammonium formate-carbonate under pressure. As an alternative a low-temperature separation method can be employed.

Small quantities of carbon monoxide needed for other purposes can be made in the laboratory in a pure condition by dehydrating pure formic acid with concentrated sulphuric acid:

$$H \cdot COOH \rightarrow CO + H_2O$$

The gas is washed free from sulphur dioxide and dried. Larger quantities can be made by reacting carbon dioxide with red-hot coke:

$$CO_2 + C \rightarrow 2CO$$

Sulphur compounds in this gas can be removed by passage over active charcoal.

SULPHUR DIOXIDE

Sulphur dioxide, although a gas at room temperature and pressure, liquefies under slight pressure. Alternatively it will liquefy at atmospheric pressure if cooled to $-10°C$. It is extremely soluble in water, only ammonia and the hydrogen halides being more so among the common gases; at 15°C and atmospheric pressure 1 volume of water dissolves 45 volumes of the gas, the solution containing some sulphurous acid, H_2SO_3.

Apart from being an intermediate in the manufacture of sulphuric acid, sulphur dioxide has certain other important industrial uses. Thus it is used as a reducing agent, and as a preservative and fumigant in agriculture, the food industry and in connection with animal and vegetable products. It is used as a refrigerant and also as a solvent in the petroleum industry.

Industrially sulphur dioxide is made by burning sulphur, although some is produced from waste gases such as smelter gases. After burning it is absorbed in water, the solution heated by low-pressure steam and the released sulphur dioxide dried by sulphuric acid and compressed into cylinders. Because of its high solubility a great deal of steam is used and some gas is lost. Consequently when recovering sulphur dioxide from waste gases it is more economic to use a solvent such as dimethylaniline. Stripping is again done by steam and drying by sulphuric acid.

Sulphur dioxide is sold as a water white liquid free from sulphur trioxide and sulphuric acid. The normal commercial grade has a purity of at least 99·9 per cent, but for refrigeration purposes an anhydrous grade with a purity of at least 99·98 per cent is sold. It is transported in steel tankers, drums or cylinders, and it can be arranged for the product to be withdrawn either as gas or as liquid.

OZONE

Ozone is a gas which is difficult to make commercially in concentrations higher than about 2 per cent in oxygen. Even at this concentration it decomposes slowly on storage. It is therefore invariably made on the site where it will be used. The method of manufacture is to pass a silent electrical discharge through air or oxygen. Although on a small scale the heat from the discharge is readily dissipated, on a larger scale positive steps must be taken to ensure its removal. The most commonly used machine is the Welsbach ozonizer which produces a product containing 1·5–2·0 per cent by weight of ozone if oxygen is used as the raw material and 1 per cent when air is used. It operates at a pressure up to 2 atm and the largest units produce 14·4 lb per hour when using oxygen.

If air is used as the raw material the product will contain not only impurities originally present in the air but also nitrous oxide and nitrogen pentoxide. The latter combines with water vapour in the air to form nitric acid. Further, if undried air is used, some hydrogen peroxide will be formed and also a lower electrical efficiency will be obtained. It is advisable therefore to use dry air. For some purposes traces of hydrogen peroxide and nitrogen pentoxide can be ignored, but if removal is necessary it can be done by scrubbing with water. When the ozone is required for chemical production it is usually made from oxygen, the unused oxygen being recycled. For other purposes such as water purification or the treatment of waste it is made from air.

It should be mentioned that ozone is a very hazardous material, the safe limit for prolonged exposure being 0·1 ppm in air compared, for example, with 100 ppm in the case of carbon monoxide. Laboratory investigations have shown that liquid ozone explodes extremely easily and so do gaseous mixtures containing over 20 per cent of ozone in oxygen.

Ozone is a powerful oxidant and a powerful germicide with no undesirable residual products as any unused product is converted to oxygen. It is used in dilute gaseous form for the purification of drinking water, for treatment of industrial wastes, for deodorizing air, for bleaching of cotton and for inhibiting growth of mould and bacteria in cold-storage chambers. It is also used in a number of organic reactions, the best known of which is the conversion of oleic acid, $C_{17}H_{33} \cdot COOH$, to equal proportions of pelargonic acid, $C_8H_{17} \cdot COOH$, and azelaic acid, $(CH_2)_7 \cdot (COOH)_2$.

The United Kingdom production of ozone is not known but in the United States it is of the order of 6000 tons per annum.

GAS MIXTURES

Users of industrial gases frequently require mixtures of two or more gases. On a very large scale, for examples mixtures of hydrogen and nitrogen or carbon monoxide, the gases are manufactured *in situ*. On a small scale suppliers of industrial gases provide many standard mixtures in cylinders. Certain mixtures are not required to be within very close tolerances and are readily and widely available. Others are specially prepared and, if requested, will be supplied with a certificate of analysis. It is important that none of the constituent gases should liquefy in the cylinder for if it did the composition of the gas withdrawn from the cylinder would not be constant. Further, the gases must not react chemically in the cylinder nor form an explosive mixture.

A number of mixtures prepared industrially have already been mentioned in this book, e.g. oxygen–helium for breathing by divers; argon–nitrogen for filling electric lamps; special mixtures for a shielding gas for welding. They are also used in semiconductor manufacture, for calibration of instruments, for leak detection, for sterilizing and for biological atmospheres.

Glossary

Absolute zero. The lowest temperature theoretically achievable.

Adiabatic process. A process taking place under conditions which do not allow heat to enter or leave the system.

Boyle's Law. The volume of a gas is inversely proportional to its pressure at constant temperature.

Cascade method. Cooling in stages using a series of coolants.

Charles' Law. The volume of a gas is proportional to its absolute temperature at constant pressure.

Critical pressure. The minimum pressure required to liquefy the gas at its critical temperature.

Critical temperature. The highest temperature at which a substance can exist in the liquid state however great the pressure.

Cryogenics. The science of low temperatures.

Entropy. A property which describes how random the molecules in a gas are. It reaches a maximum under equilibrium conditions.

Ideal gas. One which strictly obeys the laws of Boyle and Charles.

Inversion temperature. The temperature for any gas below which the Joule–Thomson effect is one of cooling and above which it is one of heating.

Joule–Thomson effect. The temperature change which occurs when a gas is allowed to expand through a narrow orifice.

Natural gas. Gas emanating from below the earth's crust, usually as a result of drilling for oil.

Perfect gas. Same as ideal gas.

Permanent gas. One whose critical temperature is below about $-60°C$.

Superconductivity. A phenomenon occurring at very low temperatures in which the resistance of a metal vanishes.

Synthesis gas. A mixture of hydrogen and carbon monoxide in proportions suitable for chemical synthesis.

Tonnage oxygen. Oxygen, often of lower purity than liquid oxygen or cylinder oxygen, made on a very large scale and usually for a single use.

Suppliers of Gases and Plants

A LARGE number of companies throughout the world supply the gases mentioned in this book. Because of the nature of the products there is very little international trade. Some of the leading companies who operate on a very large scale, and the territories in which they mainly operate, are as follows:

Air Products	United States and United Kingdom
Air Reduction Company	United States
British Oxygen Company	United Kingdom and Commonwealth
L'Air Liquide	France, Canada and Japan
Linde A.G.	Germany
Union Carbide Corporation	United States

These companies are also leaders among those undertaking to design and build low-temperature plants. While many of the plants they construct are for their own use there is also a considerable international trade in them.

Bibliography

1. Din, F. and Cockett, A. H., *Low-temperature Techniques*, Newnes, London, 1960.
2. Ruhemann, M., *The Separation of Gases*, Clarendon Press, Oxford, 1949.
3. Cremer, H. W. and Davies, T., *Chemical Engineering Practice*, Vol. 6, Butterworths, London, 1958.
4. Scott, R. B., *Cryogenic Engineering*, Van Nostrand, New York, 1959.
5. B.S. Spec. 349, Identification Colours for Gas Cylinders, 1932.
6. B.S. Spec. 4364, Specification for Industrial Oxygen, 1968.
7. Chater, W. J. B. and Harrison, J. L., *Recent Advances in Oxygen for Iron and Steel Making*, Butterworths, London, 1964.
8. Jackson, A., *Oxygen Steelmaking for Steelmakers*, Newnes, London, 1964.
9. B.S. Spec. 4365, Specification for Industrial Nitrogen, 1968.
10. Mellor, J. W., *Comprehensive Treatise on Inorganic and Theoretical Chemistry*, Vol. 8, Supplement 1 (*Nitrogen*), Longmans, London, 1964.
11. B.S. Spec. 4365, Specification for Industrial Argon, 1968.
12. Cook, G. A., *Argon, Helium and the Rare Gases*, Interscience Publishers, New York, 1961.
13. Miller, S. A., *Acetylene, Its Properties, Manufactures and Uses*, Benn, London, 1965.
14. Noyes, R., *Ammonia and Synthesis Gas*, Noyes Development Corporation, Park Ridge, New Jersey, 1967.
15. Scott, R. B., Denton, W. H. and Nicholls, C. M., *Technology and Uses of Liquid Hydrogen*, Pergamon Press, London, 1964.
16. Miller, S. A., *Ethylene and Its Industrial Derivatives*, Benn, London, 1969.
17. Haselden, G. G., *Cryogenic Fundamentals*, Academic Press, London, 1971.
18. Smith, A. U., *Current Trends in Cryobiology*, Plenum Press, London, 1970.

Index

Acetylene
  chemicals from 84–7
  production 78–9, 81–3
  properties 6, 75, 83
  storage and distribution 79–81
  for welding 84
Air
  composition 7
  purification 18–20, 35
  rectification 22–4
  separation 16 *et seq.*
Ammonia 48, 61, 91–2
Anaesthesia 74, 99, 104
Argon
  lamps 61
  production 59–61
  properties 6
  for welding 62

Butane 6,100

Calcium carbide 76–8
Carbon dioxide
  properties 6
  recovery 95–6
  storage and distribution 96–7
  uses 97–8
Carbon monoxide 6, 104–5
Cutting of metals 39–41, 101

Ethylene 6, 82, 103–4
Evaporative cooling 10, 11
Explosives 46–7, 97

Food technology 10, 53, 55–6, 97, 104, 105
Freezing mixtures 10, 11

Gas mixtures 61, 67, 107–8

Heat transfer 13–14, 20, 21, 24, 28–30, 35, 36, 55, 82
Helium
  gas uses 67
  liquid uses 68–71
  production 65–6
  properties 6
  storage 29–30
Hydrogen
  liquid 92–4
  production 88–9
  properties 6
  uses 90–1

Inversion temperature 12, 68, 92
Iron and steel manufacture 42–5

Joule–Thomson effect 12, 14, 20–1

Krypton 6, 7, 73–4

Liquefied petroleum gases 100–2

Materials of construction   24, 25, 28,
   29, 31
Medicine   45, 56, 74, 99, 104
Metal treatment   41–2, 54, 56–8, 62–3,
   97
Methane   6, 102–3

Natural gas   45, 102–3
Neon   6, 7, 72–3
Nitrogen
   chemical uses   52–3, 91–2
   inert atmospheres   53–4
   production   48–51
   properties   6, 52
   uses of liquid   54–8
Nitrous oxide   6, 99

Oxygen
   metallurgical uses   30, 39–45
   production   33–8
   properties   6, 38
   storage and distribution   26–31

Ozone   6, 8, 106–7

Propane   6, 100

Rocketry   47, 92

Storage of cryogenic liquids   13, 27–30
Sulphur dioxides   6, 105–6
Superconductivity   68–70
Synthesis gas   45–6, 89

Temperature measurement   1–2, 14–15

Xenon   6, 7, 73–4

Welding
   electric   62
   gas   39, 84, 90